T0185105

# Springer Proceedings in Mathematics & Statistics

## Volume 62

For further volumes:
http://www.springer.com/series/10533

# Springer Proceedings in Mathematics & Statistics

This book series features volumes composed of select contributions from workshops and conferences in all areas of current research in mathematics and statistics, including OR and optimization. In addition to an overall evaluation of the interest, scientific quality, and timeliness of each proposal at the hands of the publisher, individual contributions are all refereed to the high quality standards of leading journals in the field. Thus, this series provides the research community with well-edited, authoritative reports on developments in the most exciting areas of mathematical and statistical research today.

Luis F. Zuluaga • Tamás Terlaky
Editors

# Modeling and Optimization: Theory and Applications

Selected Contributions from the
MOPTA 2012 Conference

 Springer

*Editors*
Luis F. Zuluaga
Department of Industrial
    and Systems Engineering
Lehigh University
Bethlehem, PA, USA

Tamás Terlaky
Department of Industrial
    and Systems Engineering
Lehigh University
Bethlehem, PA, USA

ISSN 2194-1009          ISSN 2194-1017 (electronic)
ISBN 978-1-4939-4659-4    ISBN 978-1-4614-8987-0 (eBook)
DOI 10.1007/978-1-4614-8987-0
Springer New York Heidelberg Dordrecht London

Mathematics Subject Classification: 49-06, 49Mxx, 65Kxx, 90-06, 90Bxx, 90Cxx

Printed on acid-free paper

Springer is part of Springer Science+Business Media (www.springer.com)

# Preface

This volume contains a selection of papers that were presented at the Modeling and Optimization: Theory and Applications (MOPTA) Conference held at Lehigh University in Bethlehem, Pennsylvania, USA, between July 30 and August 1, 2012. MOPTA 2012 aimed to bring together a diverse group of researchers and practitioners, working on both theoretical and practical aspects of continuous or discrete optimization. The goal was to host presentations on the exciting developments in different areas of optimization and at the same time provide a setting for close interaction among the participants. The topics covered at MOPTA 2012 varied from algorithms for solving convex, combinatorial, nonlinear, and global optimization problems and addressed the application of optimization techniques in finance, electricity systems, healthcare, and other important fields. The five papers contained in this volume represent a sample of these topics and applications and illustrate the broad diversity of ideas discussed at the conference. Below, we briefly introduce each of them.

The paper by Anjos provides a comprehensive review of the mathematical optimization models that have been proposed to address the Unit Commitment problem. This is a fundamental problem in the operation of power systems that seeks to find the optimal way to generate a power production schedule, while ensuring demand satisfaction and the safe and reliable operation of the system. The Unit Commitment problem is becoming increasingly important and challenging. This is mainly due to the transition to low-carbon, sustainable, and renewable energy sources (e.g., wind and solar energy) and the need to reliably satisfy increasing energy demands in a scarce resource, highly competitive, and interconnected economy.

The paper by Lejeune considers a novel way to take into account uncertainty in the key problem of optimal portfolio allocation. Furthermore, the paper provides algorithmic techniques to address the solution of large-scale problems in this class with practically relevant features that make their solution more challenging. Namely, he considers the portfolio allocation problems with features such as fixed transaction costs and diversification, cardinality, and buy-in threshold constraints. Large-scale problems in this category become difficult to solve by commercially available optimization solvers. Thus, the focus is to obtain optimal or close to optimal solutions to the problem in a fast manner.

The paper by Regis considers black-box optimization problems for which the dependance between the objective function and the decision variables is not available in explicit functional form. Instead, values of the problems' objective can only be computed for given values of the decision variables through computationally expensive simulations. For this class of problems, the paper proposes an initialization strategy that can be effectively used to substantially improve the performance of solution algorithms for black-box optimization problems. This is shown by presenting corresponding computational results on problems with up to one thousand variables. In particular, instances of a black-box optimization problem arising in the management of groundwater bioremediation are considered.

The paper by Benson and Sağlam considers the solution of mixed-integer second-order cone optimization (MISOCO) problems using a combination of nonlinear, branch-and-bound, and outer approximation techniques. A key of their approach is that it allows for warmstarting when solving the continuous relaxation of the problem. The performance of their proposed techniques is investigated on MISOCO problems arising in portfolio allocation problems. Currently, MISOCO problems appear in many engineering, healthcare, and finance applications, as well as in the general context of robust optimization. Thus, this paper contributes to the development of specific algorithmic techniques for this very important class of problems.

The paper by Li and Terlaky investigates the duality relationship between two keystone algorithms to solve linear feasibility problems, namely, the perceptron and the von Neumann algorithms. This approach allows to interpret variants of the perceptron algorithm as variants of the von Neumann algorithm and vice versa and transit the complexity results from one family to the other. Advances related to this class of inexpensive algorithms are key, given the growing need to solve extremely large optimization problems in most current practical applications.

These articles address the two focus areas of the MOPTA Conference, namely, the role that modeling plays in the solution of an optimization problem and advances in optimization algorithms, theory, and applications.

We end this preface by thanking the sponsors of MOPTA 2012, namely, AIMMS (http://business.aimms.com/) , GuRoBi Optimization (http://www.gurobi.com/), IBM (http://www.research.ibm.com/), Mosek (http://www.mosek.com/), and SAS (http://www.sas.com/). We also thank the host, Lehigh University, as well as the rest of the organizing committee: Frank E. Curtis, Eugene Perevalov, Ted K. Ralphs, Katya Scheinberg, Larry V. Snyder, Robert H. Storer, and Aurélie Thiele.

Bethlehem, PA, USA                                                               Luis F. Zuluaga
                                                                                Tamás Terlaky

# Contents

# Recent Progress in Modeling Unit Commitment Problems

Miguel F. Anjos

**Abstract** The unit commitment problem is a fundamental problem in the operation of power systems. The purpose of unit commitment is to minimize the system-wide cost of power generation by finding an optimal power production schedule for each generator while ensuring that demand is met and that the system operates safely and reliably. This problem can be formulated as a mixed-integer nonlinear optimization problem, and finding global optimal solutions is important not only because of the significant operational costs but also because in competitive market environments, different near-optimal solutions can produce considerably different financial settlements. At the same time, the time available to solve the problem is a hard constraint in practice. Hence unit commitment is an important and challenging optimization problem. This article provides an introduction to the basic problem from the point of view of optimization, summarizes several related modeling developments in the recent literature while providing some possible directions for future research, and concludes with a brief mention of extensions to the basic problem that have great practical importance and are driving much current research.

**Keywords** Optimization • Unit commitment • Mixed-integer linear programming

## 1 Introduction

Electricity is a critical source of energy used by both individuals and industry on a daily basis. With increasing demand for electricity and various constraints on the development of new generation capacity, it is imperative to increase the overall efficiency of the power system. This need has motivated the concept of

M.F. Anjos (✉)
Canada Research Chair in Discrete Nonlinear Optimization in Engineering,
GERAD and École Polytechnique de Montréal, Montreal, QC, Canada H3C 3A7
e-mail: anjos@stanfordalumni.org

L.F. Zuluaga and T. Terlaky (eds.), *Modeling and Optimization: Theory and Applications*,
Springer Proceedings in Mathematics & Statistics 62, DOI 10.1007/978-1-4614-8987-0_1,
© Springer Science+Business Media New York 2013

smart grids. A smart grid is a next-generation electrical power system that combines the power distribution system with a two-way communication between suppliers and consumers to optimize the generation, transportation, and distribution of electrical energy. Smart grids are expected to deliver energy savings, cost reductions, and increased reliability and security by enabling new strategies to manage the power system. For example, a smart grid will in principle support a two-way interaction with individual homes to achieve system-side objectives [14].

Smart grids are a popular research area at present. The IEEE recently launched a new journal dedicated to it, and multiple research initiatives are underway around the world. The MOPTA 2012 conference featured five sessions of talks on electricity systems and smart grids; moreover, the AIMMS© competition problem was concerned with scheduling in smart grids.

This article is about one of the fundamental optimization problems in this area, namely, the unit commitment (UC) problem. The purpose of solving UC problems is to minimize the system-wide cost of power generation while ensuring that demand is met and that the system operates safely and reliably. The UC models were originally proposed and applied in monopolistic contexts and have been extended to generate production schedules in a competitive market environment (see, e.g., [2, 10, 21]).

Because of the operating constraints of the generators and the behavior of power flows, the complete UC problem is a mixed-integer nonlinear optimization problem. The first mixed-integer linear programming (MILP) formulation of the basic trade-offs in UC was proposed more than 50 years ago [4]. A vast literature has been dedicated to developing solution techniques for UC (see, e.g., [45, 61]). The MILP approach is among the few techniques that can provide provably global optimal UC solutions. Global optima matter in competitive markets because different near-optimal solutions may yield considerably different payments to generator owners [52].

Realistic instances of the UC problem are typically large scale and require significant computational time to solve. Moreover, in the context of system operations, the time available to solve a UC model is a hard limitation, restricting the size and scope of UC formulations. As a result, practitioners sometimes have to settle for solutions that are not globally optimal. Nevertheless, the value of the MILP approach for solving UC problems is well recognized; for example, PJM Interconnection has been using MILP since 2005 and it has contributed to important efficiency gains [43].

The objectives of this article are to provide an introduction to the UC problem from the point of view of mathematical optimization and to collect a number of recent developments in the modeling of the basic UC problem that are likely to have a significant impact on the ability to solve large-scale instances of UC in practice. The problem of UC incorporating all actual operational requirements is called *security-constrained unit commitment (SCUC)* (see, e.g., [48, Chap. 8]). Security-constrained UC includes many features that are not included in our basic UC model. We also do not explicitly address issues arising from the broader smart grid context, such as managing bidirectional flows of power, integrating renewable energy sources such as wind and solar, accounting for forecasting uncertainties, and co-optimizing generation and transmission switching. These issues are important, and because any mathematical optimization-based methodology addressing them relies on some form of the basic UC model, progress in solving the basic model will contribute to addressing them as well.

Furthermore, research on modeling basic UC remains relevant because the UC in current practice mostly uses 1-h time periods for the day-ahead planning of operations. This can be problematic in the context of competitive markets because the variability within 1 hour can be significant and may not be properly captured using 1-h periods. Hence there is a need to move to 15-min time periods in day-ahead planning. Roughly speaking this represents a fourfold increase in the complexity of the MILP models and is beyond the ability of the state-of-the-art MILP solvers even for a basic UC formulation without the additional variables and constraints in more sophisticated models.

This paper is structured as follows. In Sect. 2 we give a basic formulation of UC. In Sect. 3 we present some recent research on improving the mathematical representation of different parts of the basic formulation. In Sect. 4 we look into one of the aspects of SCUC that is not included in the basic formulation, namely the modeling of power flows in the power transmission system. Section 6 concludes the paper.

## 2   Formulating the UC Problem

Given a set of power generators and electricity demands, the basic UC problem minimizes the total power production cost so that

- the demand is met,
- sufficient reserves are in place in case of unplanned contingencies, and
- the generators operate within their physical limits:
  - the power output level of a generator is within its allowable operating range,
  - when a generator is turned on (off), it must stay on (off) for a minimum amount of time, and
  - the power output level of a generator cannot change too rapidly.

The UC problem is usually defined over a planning period ranging from 1 day (24 h) to 1 week (168 h). We consider the point of view of the system operator that is given a forecast of the total hourly demand for the planning period and needs to decide which generating units will be committed (i.e., running) and which will be de-committed (i.e., not running) during each hour. This planning is necessary because demand varies from hour to hour and from one day to another, and since different units have different generation costs, starting up and shutting down units suitably often results in significantly lower costs. Moreover, the integration of wind generation is also leading to an increase in start-ups and shutdowns of units [54].

The remainder of this section describes a basic MILP formulation of UC. It provides a good starting description of the UC problem. This formulation is largely based on Bhattacharya et al. [8, Chap. 2, Sect. 4] and Arroyo and Conejo [2].

## 2.1  Costs

**Fuel Cost**

For conventional generators such as thermal units, the fuel cost $c_g^f$ of generator $g$ is typically represented by a quadratic function of the form

$$c_g^f(p) := a_g p^2 + b_g p + c_g, \tag{1}$$

where $p$ denotes the number of megawatts (MW) generated. It is frequently assumed that the function $c_g^f$ is strictly convex since the marginal cost of power increases as a function of $p$. However, this is not always the case. For instance, in the case of combined cycle gas turbines, steam admission valves are opened one after the other when increasing the power output; the result is a nonconvex (but piecewise convex) variable cost function.

For MILP models the quadratic function is typically approximated by a piecewise linear function. This is the subject of Sect. 3.1.

**Start-Up and Shutdown Costs**

The start-up cost is typically modeled as a function of the length of time for which the unit was offline:

$$c_g^{su}(t) := c_g^{su} + b_g^{su}\left(1 - e^{-t/d_g^{cool}}\right),$$

where $c_g^{su}$ is a fixed cost for start-up, $b_g^{su}$ is the additional cost of a cold start-up, $t$ is the length of time for which the unit has been off, and $d_g^{cool}$ is a constant representing the cooling speed of the unit.

For the MILP models this form of the start-up cost is often discretized using a step function that is constant for each time period. Let $T$ be the set of time periods in the planning period, and suppose for ease of presentation that the step function has a separate segment for each time period $t \in T$. The step function may then be represented in the form

$$\hat{c}_g^{su}(t) = c_g^{su} + \sum_{i=1}^{T} \hat{b}_g^{su}(i)\beta_g(i,t), \forall\, t \in T, \tag{2}$$

where $\beta_g(i, t)$ is a binary variable that equals 1 if unit $g$ is started up at time period $t$ and has been off for $i$ time periods, and each constant $\hat{b}_g^{su}(i)$ equals the value of $b_g^{su}\left(1 - e^{-t/d_g^{cool}}\right)$ with $t$ set to the appropriate value to reflect the fact that the unit has been off for $i$ time periods.

Shutdown costs are usually not significant in comparison to the other costs and thus often ignored. If included, a constant cost $c_g^{sd}$ is usually assumed.

## 2.2 Variables

The (continuous) power variables are

$$p_g(t) = \text{power (MW) scheduled to be produced by generator } g \text{ at time } t.$$

We also have three sets of binary variables to model the status of generator $g$:

$$s_g(t) = \begin{cases} 1, & \text{if generator } g \text{ is on at time period } t, \\ 0, & \text{otherwise}, \end{cases}$$

$$s_g^{su}(t) = \begin{cases} 1, & \text{if generator } g \text{ starts up at time period } t, \\ 0, & \text{otherwise}, \end{cases}$$

$$s_g^{sd}(t) = \begin{cases} 1, & \text{if generator } g \text{ shuts down at time period } t, \\ 0, & \text{otherwise}. \end{cases}$$

The three sets of variables are logically linked: $s_g^{su}(t_1) = 1$ means that $s_g(t_1 - 1) = 0$ and $s_g(t_1) = 1$, i.e., at time period $t_1$, unit $g$ starts up and the variable $s_g$ switches from 0 to 1. It will then remain at 1 until $s_g^{sd}(t_2) = 1$ for some $t_2 > t_1$, and thus $s_g(t_2 - 1) = 1$, $s_g(t_2) = 0$, and generator $g$ is shut down from time period $t_2$ onwards until it is started up again.

## 2.3 Objective Function

Based on the quantities introduced above, the objective function of UC is

$$\sum_{t \in T} \sum_{g \in G} c_g^f(p_g(t)) \, s_g(t) + c_g^{su} s_g^{su}(t) + c_g^{sd} s_g^{sd}(t), \tag{3}$$

where $G$ is the set of generators.

## 2.4  Constraints

### Supply-Demand and Spinning Reserves

These constraints that sufficient generation is scheduled in order to satisfy the total demand as well as the spinning reserves required:

$$\sum_{g \in G} p_g(t) \geq D(t) + R(t), \forall t \in T, \tag{4}$$

where $D(t)$ and $R(t)$ are respectively the demand (MW) and spinning reserve (MW) at time period $t$.

The spinning reserve is an amount of generation capacity available to the system operator over and above the forecasted demand. It is called spinning because it is available almost instantaneously in case of need. The reserve is typically set as the sum of (i) a base amount, (ii) a fraction of the demand, and (iii) a fraction of the maximum power output of the largest unit turned on [48].

### Generation Limits

The power output of a generator $g$ that is turned on during time period $t$ must be within the upper and lower limits of its allowable operating range:

$$\underline{P}_g s_g(t) \leq p_g(t) \leq \overline{P}_g s_g(t) \ \forall g \in G, \ \forall t \in T, \tag{5}$$

where $\underline{P}_g$ and $\overline{P}_g$ are in MW and denote respectively the minimum and maximum power output levels of generator $g$.

If unit $g$ is off during time period $t$, these constraints set $p_g(t)$ to zero.

### Minimum Uptime and Downtime

The minimum uptime constraints ensure that a unit $g$ that is turned on will not be turned off before it has been on for at least its minimum uptime of $UT_g$ time periods:

$$\sum_{i=t}^{t+UT_g-1} s_g(i) \geq UT_g s_g^{su}(t), \quad \forall t = L_g + 1, \ldots, |T| - UT_g + 1, \ \forall g \in G \tag{6}$$

$$\sum_{i=t}^{|T|} (s_g(i) - s_g^{su}(t)) \geq 0, \quad \forall t = |T| - UT_g + 1, \ldots, |T|, \ \forall g \in G, \tag{7}$$

where $L_g = \min\{|T|, U_g\}$ and $U_g$ is the number of time periods that unit $g$ is required to be on at the start of the planning period.

Similarly, the minimum downtime constraints ensure that a unit $g$ that is turned off will not be turned on before it has been off for at least its minimum downtime of $DT_g$ time periods:

$$\sum_{i=t}^{t+DT_g-1} (1 - s_g(i)) \geq DT_g s_g^{sd}(t), \quad \forall t = F_g 1, \ldots, |T| - DT_g + 1, \forall g \in G,$$
(8)

$$\sum_{i=t}^{|T|} (1 - s_g(i) - s_g^{sd}(t)) \geq 0, \quad \forall t = |T| - DT_g + 2, \ldots, |T|, \forall g \in G,$$ (9)

where $F_g = \min\{|T|, D_g\}$ and $D_g$ is the number of time periods that unit $g$ is required to be off at the start of the planning period.

Also at the start of the planning period, there may be some generators that have recently been turned on or turned off. Equations (11) and (10) force these generators to remain on and off for the appropriate amount of time to satisfy the minimum up- and downtime constraints:

$$\sum_{t=1}^{L_g} s_g(t) = L_g, \forall g \in G,$$
(10)

$$\sum_{t=1}^{F_g} s_g(t) = 0, \forall g \in G.$$
(11)

The effect of these constraints is to correctly fix the variables $s_g(t)$ to 0 or 1.

## Ramping Rates

The power output of a generator, particularly that of a coal-based thermal unit, cannot fluctuate too rapidly. Thus the change in the value of $p_g(t)$ is constrained by specified ramp-up and ramp-down rates and start-up and shutdown rates.

The ramp-up constraints (12) limit the increase in power output allowed between consecutive time periods. At a given time period $t$, if the generator was on in the previous time period ($s_g(t-1) = 1$), then the increase in power output cannot be larger than the ramp-up rate $RU_g$. Moreover, if the generator is turned on at $t$ (i.e., if $s_g^{su}(t) = 1$), then $s_g(t-1) = 0$ and the generator can produce at most $SU_g$ in time period $t$, where $SU_g$ is the maximum start-up rate of generator $g$. This yields the ramp-up constraints:

$$p_g(t) - p_g(t-1) \leq RU_g s_g(t-1) + SU_g s_g^{su}(t), \quad \forall g \in G, \forall t \in T,$$ (12)

where $RU_g$ is the maximum ramp-up rate (in MW) of generator $g$ and $SU_g$ is the maximum start-up rate (in MW) of generator $g$.

A similar argument yields the ramp-down constraints:

$$p_g(t-1) - p_g(t) \leq RD_g s_g(t) + SD_g s_g^{sd}(t), \quad \forall g \in G, \ \forall t \in T, \quad (13)$$

where $RD_g$ is the maximum ramp-down rate (in MW) of generator $g$ and $SD_g$ is the maximum shutdown rate (in MW) of generator $g$.

## 3   Recent Modeling Improvements

In this section we present a number of recent ideas for improving the mathematical representation of the various components of the basic UC formulation.

### 3.1   Fuel Cost

Recall that for MILP modeling we want to approximate the quadratic function (1) by a piecewise linear function. Let us denote this approximation as

$$\hat{c}_g^f(p) = c_g + \sum_{\lambda=1}^{\Lambda} \hat{b}_g^\lambda p, \quad (14)$$

where $\lambda$ indexes the segments of the piecewise approximation of $c_g^f$.

It is always possible to improve the approximation by increasing the number $\Lambda$ of linear segments. This will however increase the number of binary variables of the model which may be problematic for large-scale formulations, such as those with 15-min time periods in day-ahead UC. We therefore look at ways of improving the quality of the approximation for a fixed number $\Lambda$ of line segments.

#### Optimal Interval Partition

Let us first consider the case where no convexity assumption is made about $c_g^f(p)$, i.e., the sign of $a_g$ is arbitrary. Wu [58] proposes an approach that optimizes the location of the breakpoints defining the line segments within the interval of power operating levels allowed. Specifically, the aim is to minimize the difference between the arclength of $c_g^f(p)$ and the arclength of $\hat{c}_g^f(p)$ over the interval.

Let $p^{(1)}, p^{(2)}, \ldots, p^{(\Lambda-1)}$ be the $\Lambda - 1$ breakpoints to place between the (fixed) endpoints of the interval $p^{(0)} = \underline{P}_g$ and $p^{(\Lambda)} = \overline{P}_g$. The arclength of $c_g^f(p)$ between the consecutive breakpoints $p^{(\lambda-1)}$ and $p^{(\lambda)}$ is

$$a_\lambda^{c_g^f} = \int_{p^{(\lambda-1)}}^{p^{(\lambda)}} \sqrt{1 + \left(\frac{\partial c_g(p)}{\partial p}\right)^2}\, dp$$

$$= \int_{p^{(\lambda-1)}}^{p^{(\lambda)}} \sqrt{1 + (2a_g p + b_g)^2}\, dp$$

$$= \frac{1}{4a_g} \left[ (2a_g p + b_g)\sqrt{1 + (2a_g p + b_g)^2} + \ln\left((2a_g p + b_g)\sqrt{1 + (2a_g p + b_g)^2}\right) \right]_{p^{(\lambda-1)}}^{p^{(\lambda)}}$$

while the length of the corresponding line segment of $\hat{c}_g^f(p)$ is

$$a_\lambda^{\hat{c}_g^f} = \sqrt{\left((a_g(p^{(\lambda)})^2 + b_g p^{(\lambda)} + c_g) - (a_g(p^{(\lambda-1)})^2 + b_g p^{(\lambda-1)} + c_g)\right)^2 + \left(p^{(\lambda)} - p^{(\lambda-1)}\right)^2}.$$

Thus the optimization problem to place the breakpoints is formulated in [58] as

$$\min_{p^{(\lambda)}, \lambda=1,\dots,\Lambda-1} \sum_{\lambda=1}^{\Lambda} \sqrt{\left(a_\lambda^{c_g^f}\right)^2 - \left(a_\lambda^{\hat{c}_g^f}\right)^2}. \tag{15}$$

The KKT optimality conditions for this unconstrained optimization problem are

$$\frac{\partial}{\partial p^{(\lambda)}} \left[ \sum_{\lambda=1}^{\Lambda} \sqrt{\left(a_\lambda^{c_g^f}\right)^2 - \left(a_\lambda^{\hat{c}_g^f}\right)^2} \right] = 0, \quad \lambda = 1, \dots, \Lambda - 1. \tag{16}$$

This is a system of $\Lambda - 1$ equations in $\Lambda - 1$ unknowns, albeit nonlinear. It can be solved by one of several well-known optimization methods for such problems (see, e.g., [15]). Since this problem is nonconvex, there remains the issue of the starting point. An obvious choice for the starting point would be equispaced breakpoints, but it would be interesting to study the sensitivity of the solution to the choice of starting point.

More generally, if we define the vectors $A^{c_g^f} = \left(a_\lambda^{c_g^f}\right)_\lambda$ and $A^{\hat{c}_g^f} = \left(a_\lambda^{\hat{c}_g^f}\right)_\lambda$, then the optimization problem can be set up as

$$\min_{p^{(\lambda)}, \lambda=1,\dots,\Lambda-1} \|A^{c_g^f} - A^{\hat{c}_g^f}\|$$

for some choice of norm $\|\cdot\|$. It is likely that the optimal breakpoints will depend on the choice of the norm. This is another question that could be investigated further.

**Perspective Cuts**

We now assume that $c_g^f(p)$ is strictly convex, i.e., that $a_g > 0$ holds. We further assume that the breakpoints for constructing the piecewise linear approximation $\hat{c}_g^f$ have been determined, e.g., using the approach described in the previous subsection. A common way to build $\hat{c}_g^f$ is to have it coincide with $c_g^f$ at the $\Lambda + 1$ breakpoints. This gives a convex upper approximation of $c_g^f(p)$. Different constructions have been proposed, but most often they are defined in the space of the $p$ variables and they do not involve the $s_g(t)$ variables.

Frangioni et al. [17] propose a construction of the piecewise linear approximation where each term $c_g^f(p_g(t))s_g(t)$ in (3) is replaced by a new variable $\gamma_g$ and the following $\Lambda + 1$ constraints are added to the formulation:

$$\gamma_g \geq (2a_g p^{(\lambda)} + b_g)p_g(t) + (c_g - a_g(p^{(\lambda)})^2)s_g(t), \quad \lambda = 0, 1, \ldots, \Lambda. \quad (17)$$

This is the *perspective-cut* approximation.

The perspective-cut approximation is motivated by the theory developed in [16] where it is proved that the constraints (17) are supporting hyperplanes to the graph of the convex envelope of the function $a_g p^2 + b_g p + c_g s_g(t)$. This approximation is, in this sense, the best possible convex approximation of the term $c_g^f(p_g(t))s_g(t)$. Since this function is equal to $c_g^f(p_g(t))s_g(t)$ at every binary feasible solution, this means that the approximation is actually exact at binary feasible solutions. Moreover, it does not overestimate the cost function as an upper approximation does.

The perspective-cut approach can be applied to cost functions that are nonconvex but piecewise convex by approximating each convex segment independently. This includes for example cost functions with valve points as arise in modeling combined cycle gas turbines.

The computational results reported in [17] suggest that the perspective-cut approach incorporated into a basic formulation of UC yields comparable or slightly better solutions in less time than if a standard piecewise linear approximation is used. Moreover, the perspective-cut approach can be applied in a dynamic fashion where additional constraints of the form (17) are generated iteratively as needed. According to [17], such a dynamic approach, if appropriately controlled, can provide even better results.

## 3.2   *Upper Generation Limit*

Ostrowski et al. [41] introduced the following modified form of the upper bound in (5) to make the upper bound depend on $s_g^{su}$ and $s_g^{sd}$:

$$p_g(t) \leq \overline{P}_g s_g(t + K_g(t)) + \sum_{i=1}^{K_g(t)} (SD_g + (i-1)RD_g)s_g^{sd}(t+i) - \sum_{i=1}^{K_g(t)} \overline{P}_g s_g^{su}(t+i) \; \forall \, t \in T,$$

$$(18)$$

where $K_g(t) = \max\{k \in \{1, \ldots, UT_g\}|SD_g + (k-1)RD_g < \overline{P}_g$ and $k+t < |T|\}$. This tighter upper bound follows from the following observations:

- If a given generator is turned off at time $t + 1$, then the generator cannot produce more than $SD_g$ in time $t$. Similarly, if the generator is turned off at time $t + 2$, then it cannot produce more than $SD_g + RD_g$ in time $t$ because it needs to be able to ramp down to $SD_g$ or below in time $t + 1$. On the other hand, if $k$ is such that $SD_g + (k - 1)RD_g \geq \overline{P}_g$, then knowing the generator will be turned off at time $t + k$ does not affect the upper production limits at time $t$ because even if the generator is producing at maximum capacity at time $t$, there is sufficient time for the generator to ramp down and shut off at time $t + k$.
- If the generator is not turned off in the interval $[t + 1, t + K_g(t)]$, then the upper bound for $p_g(t)$ is $\overline{P}_g$ if the generator is on at time $t$ and 0 if it is off. If the generator is on at $t$ and is not turned off in the interval, then all variables on the right-hand side of (18) are 0 except the $s_g$ term, making the right-hand side $\overline{P}_g$. If the generator is off at time $t$, then either all variables on the right-hand side of (18) are 0 or one of the $s_g^{su}$ terms and the $s_g$ term are equal to 1. These terms cancel each other out, making the right-hand side equal to 0.

The term $K_g(t)$ is bounded above by $UT_g$ to ensure that only one $s_g^{sd}$ term takes the value 1 in the interval $[t + 1, t + K_g(t)]$. The computational results in [41] report on the use of (18) in combination with the improved ramping cuts (21)–(25) in Sect. 3.4 below. The results show that the combined set of constraints can have a significant impact on the quality of the UC formulation and the computational time required to solve it.

## 3.3 Minimum Uptime and Downtime

Rajan and Takriti [46] proposed an alternative formulation for the minimum uptime and downtime constraints. These inequalities are called the *turn-on* and *turn-off* inequalities. The turn-on inequalities are concerned with the minimum uptime:

$$\sum_{k=t-UT_g+1, \, k \geq 1}^{t} s_g^{su}(k) \leq s_g(t) \quad \forall t \in [L_g + 1, \ldots, |T|] \, \forall g \in G, \quad (19)$$

while the turn-off inequalities are concerned with the minimum downtime:

$$s_g(t) + \sum_{k=t-DT_g+1, \, k \geq 1}^{t} s_g^{sd}(k) \leq 1 \quad \forall t \in [F_g + 1, \ldots, |T|] \, \forall g \in G. \quad (20)$$

These inequalities arise from the work of Lee et al. [30] that proposed another class of inequalities called the alternating up/down inequalities. Rajan and Takriti [46] proved that the inequalities (19) and (20) dominate the alternating up/down inequalities, i.e., that any fractional solution in the LP relaxation that satisfies the constraints (19) and (20) will also satisfy the alternating up/down inequalities. The computational results in [46] as well as the subsequent papers [38, 41] have confirmed that this new set of minimum up/downtime inequalities can have a significant impact on the computational time needed to solve UC problems.

## 3.4  Ramping Constraints

### General Ramping Constraints

The ramping constraints can be strengthened under certain circumstances. Ostrowski et al. [41] proposed the following tightened form of the ramp-down inequalities:

If $RD_g > (SU_g - \underline{P}_g)$ and $UT_g \geq 2$, then for $t = 1, \ldots, |T|$

$$p_g(t-1) - p_g(t) \leq RD_g s_g(t) + SD_g s_g^{sd}(t) - (RD_g - SU_g + \underline{P}_g) s_g^{su}(t-1)$$

$$- (RD_g + \underline{P}_g) s_g^{su}(t). \tag{21}$$

If generator $g$ is turned on at time $t - 1$, the difference between the start-up rate $SU_g$ and the minimum power limits $\underline{P}_g$ may provide a tighter bound than the ramp-down rate. This is because $p_g(t - 1)$ must be below $SU_g$ and $p_g(t)$ must be greater than $\underline{P}_g$. Since both $s_g(t)$ and $s_g^{su}(t - 1)$ equal 1 in this case, and under the assumption that $RD_g > (SU_g - \underline{P}_g)$, the right-hand side $SU_g - \underline{P}_g$ is a tighter constraint than the ramping rate $RD_g$. The constraints (21) are further strengthened by forcing $p_g(t - 1) - p_g(t)$ to be negative if the generator is turned on at time $t$. In this case, the constraint enforces that $p_g(t) \geq \underline{P}_g$.

The assumption $RD_g > (SU_g - \underline{P}_g)$ forces the term $-(RD_g - SU_g + \underline{P}_g) s_g^{su}(t-1) - (RD_g + \underline{P}_g) s_g^{su}(t)$ to always be negative. As a consequence, the set of constraints (21) will always improve on (13), i.e., any fractional solution in the LP relaxation that satisfies the constraints (21) will also satisfy (13). Thus the constraints (21) dominate the ramp-down constraints (13), and if constraints (21) are included in the formulation, constraints (13) can be omitted.

Similarly, we have the set of constraints (22) that additionally take into account information from time $t + 1$:

If $RD_g > (SU_g - \underline{P}_g)$, $UT_g \geq 3$, and $DT_g \geq 2$, then for $t = 1, \ldots, |T|$

$$p_g(t-1) - p_g(t) \leq RD_g s_g(t+1) + SD_g s_g^{sd}(t) + RD_g s_g^{sd}(t+1)$$

$$- (RD_g - SU_g + \underline{P}_g) s_g^{su}(t-1) - (RD_g + \underline{P}_g) s_g^{su}(t) - RD_g s_g^{su}(t+1). \tag{22}$$

One more strengthened constraint comes from bounding ramping over two time periods:

For $t = 2, \ldots, |T|$

$$p_g(t - 2) - p_g(t) \leq 2RD_g s_g(t) + SD_g s_g^{sd}(t - 1) + (SD_g + RD_g)s_g^{sd}(t)$$
$$-2RD_g s_g^{su}(t - 2) - (2RD_g + \underline{P}_g)s_g^{su}(t - 1) - (2RD_g + \underline{P}_g)s_g^{su}(t) \tag{23}$$

For the ramp-up constraints, we have constraint (24) that is similar to (21):

If $RU_g > (SD_g - \underline{P}_g)$ and $UT_g \geq 2$, then for $t = 1, \ldots, |T|$

$$p_g(t) - p_g(t - 1) \leq RU_g s_g(t) - \underline{P}_g s_g^{sd}(t) - (RU_g - SD_g + \underline{P}_g)s_g^{sd}(t + 1)$$
$$+ (SU_g - RU_g)s_g^{su}(t). \tag{24}$$

If $SD_g - \underline{P}_g < RU_g$ and the generator is turned off in time $t + 1$, then the upper bound of $p_g(t) - p_g(t-1)$ provided by the $SD_g - \underline{P}_g$ is stronger than the upper bound provided by $RU_g$. Furthermore, the addition of the $s_g^{sd}(t)$ term forces the right-hand side to be negative if the generator is turned off at time $t$ (implying $p_g(t-1) \geq \underline{P}_g$). These constraints (24) dominate the constraints (12).

One more set of strengthened ramp-up constraints bounds ramping up over two time periods (analogously to (23) for ramping down):

If $RU_g > (SD_g - \underline{P}_g)$, $UT_g \geq 2$, and $DT_g \geq 2$, then for $t = 2, \ldots, |T|$

$$p_g(t) - p_g(t - 2) \leq 2RU_g s_g(t) - \underline{P}_g s_g^{sd}(t - 1) - \underline{P}_g s_g^{sd}(t)$$
$$+ (SU_g - RU_g)s_g^{su}(t - 1) + (SU_g - 2RU_g)s_g^{su}(t). \tag{25}$$

Ostrowski et al. [41] proved that under minimal assumptions, the strengthened ramping inequalities (21)–(25) are facets for specific projections of the convex hull of binary feasible solutions of the UC problem. The computational results in [41] demonstrate that these strengthened constraints can have a significant impact on the quality of the UC formulation and the computational time required to solve it.

## Start-Up and Shutdown Ramping

A different strategy to tighten the ramping constraints is proposed in the very recent paper of Morales-España et al. [38]. In this paper the authors focus on the distinction between the ramping during start-up and shutdown and the operational ramping done when the unit is operating above the minimum production level. While theirs is not the first MILP formulation of these constraints (see [3, 51]), the contribution in [38] is an improved formulation of the start-up and shutdown ramping constraints that improves the efficiency of the MILP approach without increasing the number of constraints or the number of variables. We summarize here the key ideas behind their proposed formulation of the start-up and shutdown ramping constraints. The issue of the number of binary variables in MILP formulations is addressed in Sect. 3.5.

In contrast to the models mentioned so far, the approach of [38] uses binary commitment variables that identify whether the unit is up/down (as opposed to online/offline) in the sense that the generator $g$ is up if its power output is at or above $\underline{P}_g$ and down if its output is below $\underline{P}_g$. The up and down states are handled differently:

- when a unit is up, its power output can follow any trajectory that respects the generation limits and general ramping constraints,while
- when it is down, it is either starting up or shutting down, and in either case its output follows a predefined trajectory.

The formulation makes use of variables modeling the different start-up types. These types are defined by the time that each unit has been down. Specifically, the start-up type $\delta_{t,\ell}$ is selected if at time period $t$ the unit has been down within the $\ell$th time interval, and the time intervals are defined based on the step function approximation of the exponential start-up cost described in Sect. 2.1. Additional constraints are used to ensure a unit of the hottest type is selected when a start-up is required. It is worth noting that while the $\delta_{t,\ell}$ variables are binary by definition, the nature of the start-up costs ensures that they will take a binary value at optimality even if they are defined as continuous variables in the model.

## 3.5  Fewer or More Binary Variables?

It is straightforward to verify that the values of $s_g^{\text{su}}$ and $s_g^{\text{sd}}$ can be determined from the values of $s_g$. Carrion and Arroyo demonstrated that it is possible to use this observation to obtain an *efficient formulation* of the UC problem using only the $v$ variables [12]. The advantage of this method is that the number of binary variables is reduced by a factor of three.

Morales-España et al. [38] also seek a formulation with fewer variables by omitting the power output variables as well as the online/offline variables and instead computing the energy production of each unit. It is then possible to compute the optimal values of the omitted variables from the optimal solution to the formulation.

Both [12, 38] report results indicating that these changes lead to a significant reduction in the computation time. However, it is not yet well understood whether a reduced number of binary variables is always beneficial. In particular, while the computational results reported in [12] indicate that the efficient formulation can be beneficial, the results in [41] seem to contradict this. The conclusions that can be drawn from a direct comparison of the results in these two papers are limited by the fact that they used different (but similarly generated) data sets and significantly different versions of the CPLEX solver.

Nevertheless some general observations can be made [41]. While somewhat counterintuitive, it is possible that formulations with more binary variables are easier to solve. One reason is that it is not always true that when fewer variables are relaxed from binary to continuous, fewer variables will be branched on, and the MILP solution tree will thus be smaller. For instance, it may happen that branching on a $s_g^{su}$ or $s_g^{sd}$ variable is a better choice than any of the $s_g$ variables when it comes to the size of the resulting MILP tree. While making the "best" branching decision is very difficult in general (even for sophisticated solvers), if an MILP solver is reasonably adept at picking branching variables, a larger pool of branching candidates should produce a smaller enumeration tree. A second reason is that generating strong cutting planes may be easier when more binary variables are present. If the solver does not know that certain quantities are inherently binary, it is not able to look for opportunities to exploit that fact.

Thus it may be profitable to consider more sophisticated approaches. For example, a strategy that partially uses the insight behind the efficient formulation is to give priority for branching to the $s_g$ variables over the $s_g^{su}$ and $s_g^{sd}$ variables. This ensures that the latter are chosen only after all $s_g$ variables have been fixed and maintains the advantages of the presence of the $s_g^{su}$ and $s_g^{sd}$ variables for generating cutting planes.

It is an interesting topic for future research to explore the importance of the various groups of binary variables and thus develop branching rules specifically for large-scale UC problems.

## 3.6 Symmetry in the Presence of Multiple Identical Units

Instances of UC with multiple identical generators can be surprisingly difficult to solve. This is because the presence of multiple generators with identical character- istics introduces many symmetries into the MILP problem. It is well known that the presence of symmetry in MILP can cause serious computational difficulties. Nearly 40 years ago Jeroslow [22] presented a class of simple integer programming problems that require an exponential number of subproblems when using pure branch and bound due to the problem's large symmetry group.

Symmetry-breaking constraints have been used to improve formulations of telecommunication problems, noise pollution problems, and others (see, e.g., [49]). The past decade has seen advances in general methods for symmetry breaking, most notably isomorphism pruning [34, 35] and orbital branching [42]. These methods are advantageous for problems with general symmetry groups. There are important classes of MILP problems, such as bin packing and graph coloring, with highly structured symmetry groups. This observation has motivated the development of problem-specific techniques. For example, orbital fixing [25, 26] is an efficient way to break symmetry in set partitioning problems.

While we know ways to break symmetry, we do not yet have a clear understand- ing of the most effective way to break symmetry in particular contexts and, more importantly, of how symmetry-breaking methods interact with other MILP features such as branching strategies and cutting plane methods. Since UC problems exhibit a special symmetric structure in the presence of multiple generators with identical characteristics, and since MILP is the current industry standard for solving UC problems, effective symmetry-breaking methods specifically tailored for UC are of interest.

One way to remove symmetry from the UC problem's formulation is to aggregate all identical generators into a single generator. This is typically done for combined cycle plants that can have two or three identical combustion turbines. A comparison between aggregating combined cycle generators and modeling them as individual units was made in [32]. While this might be effective in reducing the number of variables, aggregating generator variables may not be straightforward, and some of the physical requirements may be difficult to enforce. If the UC solution only gives the number of generators operating at each time period as well as the total power produced by those generators, it may be difficult to determine how much power each unit produces in each time period.

Alternatively it is possible to exploit a new technique called modified orbital branching (modOB). Ostrowski et al. [40] proposed modOB as a means to solve more efficiently certain problems with structured symmetry groups. For example, modOB solves the Jeroslow problem with a single branch. This improvement is particularly important for problems that can be expressed using orbitopes, as is the case for the UC problem.

To briefly explain the concept of an orbitope, consider the graph coloring problem. Recall that feasible solutions for graph coloring are equivalent under color permutations. Let us express the possible colorings of a graph using 0/1 matrices such that entry $i, j$ equals 1 if and only if node $i$ is assigned color $j$. Under this representation, color permutations correspond to column permutations of the solution matrix. Hence the symmetry acting on the (matrix) variables consists of all permutations of the columns. A common technique used to remove equivalent solutions to the graph coloring problem is to restrict the feasible region to matrices with lexicographically decreasing columns. This idea can be generalized to other problems with this structure. The convex hull of all $m \times n$ 0/1 matrices with lexicographically decreasing columns is called a *full orbitope*. A complete description of full orbitopes is not known, though an extended formulation is given in [24].

Since the variables in UC can be expressed as full orbitopes, modOB can be applied directly in the UC context. Ostrowski et al. [40] showed how to use modOB to restrict the branch-and-bound search to solutions in the full orbitope. To provide some insight on how modOB works in the UC context, suppose that we obtain the following (partial) solution for one of the subproblems in the enumeration tree:

$$
x = \begin{pmatrix} 1 \ ? \ ? \ ? \ ? \\ ? \ ? \ ? \ 1 \ ? \\ ? \ 1 \ ? \ ? \ ? \\ ? \ 1 \ 1 \ ? \ ? \end{pmatrix}.
$$

In the subproblem, all of the column permutations have been removed from the symmetry group. If

$$
x_{LP} = \begin{pmatrix} 1 & 0.95 & 1 & ? & ? \\ ? & ? & ? & 1 & ? \\ 0.97 & 1 & 1 & ? & ? \\ 0.95 & 1 & 1 & ? & ? \end{pmatrix}
$$

is the corresponding LP solution, then, while technically there is no symmetry in this subproblem, it is likely that the optimal solution to this problem has each of $x_{1,2}$, $x_{3,1}$, and $x_{4,1}$ equal to 1. If these variables had been fixed, then all permutations of the first three columns of $x$ would be equivalent, and modOB could exploit this symmetry to strengthen the branching.

One issue with using modOB in practice is that MILP solvers will generally *not* choose to branch variables with a value close to 1 in the LP solution (and especially not if they are equal to one) because doing so will not improve the bound. However, from a symmetry point of view, those variables should be chosen and fixed to strengthen the subsequent branches. The approach taken in [40] is to first let CPLEX choose the branching candidate, then that branching decision is augmented using modOB so that there is a better chance that useful symmetry will be recognized and exploited. This strategy could in principle be used in conjunction with other MILP solvers.

The computational results reported in [40] suggest that modOB performs in general significantly better than the original orbital branching, sometimes reducing the running time by as much as one order of magnitude. Unfortunately it was not possible to make a direct comparison between CPLEX with modOB and CPLEX without modOB because all versions of orbital branching were implemented using the branch callback feature, and this disables other CPLEX features that also impact CPLEX's performance on UC instances. It would be very interesting to quantify the impact of modOB if it were incorporated into full default CPLEX.

## 4 Power Flows and the Transmission System

The basic formulation of UC in Sect. 2 does not account for a number of important aspects of SCUC. Among them are transmission limits on the power flow across a line, reactive power requirements (generation limits as well as reserves) for system stability, and bus voltage limits. In other words, while the basic model allocates sufficient generation for demand and spinning reserves, it does not explicitly check that the power can suitably flow within the system. The effect of the transmission system is particularly important for UC in competitive markets where issues like the congestion of the power lines and the different locational marginal prices of electricity are essential to support reliable market operation.

Because the AC power flow equations are nonlinear and nonconvex, most UC models either ignore the transmission system or use a linear approximation of it. The solutions obtained are then adjusted as necessary to ensure their feasibility with respect to the neglected quantities such as voltage and reactive power. Under normal conditions, this approach provides a reasonably good estimate of the neglected components. However, under extreme scenarios, it can lead to misleading results, for example, in quantifying the effect of market power [5, 6]. Thus, UC models that account accurately for the impact of the transmission system are essential, particularly in competitive market environments. This section looks at existing techniques and promising directions for incorporating the AC power flow into UC models.

Significant work has been carried out by Shahidehpour and coauthors using approaches based on solving MILP models using Benders decomposition [18, 19, 33]. The basic idea is to check compliance with the AC power flow equations in the Benders subproblem. If the flow cannot converge, or if violations of transmission flows and/or bus voltages are found, then a corresponding Benders cut is computed and added to the master problem. In this way, the Benders cuts are the means to couple the active and reactive power adjustments.

In the paper by Fu et al. [18], the authors found that it pays off to initially solve the problem with a (linear) DC representation of the system to obtain a good initial estimate of the system state (including bus voltages and reactive power levels) before launching the Benders algorithm on the AC problem. Since this work is set in the context of the complete SCUC problem, it is important to mention that it considers not only the necessary constraints on transmission flows and bus voltages but also the interaction between active and reactive power in a market operation framework. Extensions of this approach are presented in [19, 33]. The computational results support the conclusion that the Benders-based approach is an efficient tool, and that the results provide a satisfactory trade-off between technical security, economical realities, and computation time for real-time applicability.

Beyond the literature on UC, there is a growing body of literature on the linearization of the AC power flow equations. These can be grouped into two broad classes of methods, namely iterative methods and declarative models [13, Section VI]. While iterative methods are generally computationally efficient, from our perspective they suffer the drawback that they cannot be easily integrated into an MILP or similar declarative approach. For this reason, we focus our attention on declarative modeling approaches as these may be particularly amenable to incorporation in the basic UC model.

Let us state a commonly used form of the AC power flow equations and the corresponding DC linearization. For a line from bus $n$ to bus $m$, we assume that we are given the line's series susceptance $B_{nm}$ and its series conductance $G_{nm}$. Then the AC power flow equations may be expressed in the form

$$P_{nm} = |V_n|^2 G_{nm} - |V_n||V_m|G_{nm}\cos(\theta_n - \theta_m) - |V_n||V_m|B_{nm}\sin(\theta_n - \theta_m),$$
(26)

$$Q_{nm} = -|V_n|^2 B_{nm} + |V_n||V_m|B_{nm}\cos(\theta_n - \theta_m) - |V_n||V_m|G_{nm}\sin(\theta_n - \theta_m),$$
(27)

where $P_{nm}$ and $Q_{nm}$ are respectively the active and reactive power flow from bus $n$ to bus $m$ in per unit (p.u.), $|V_n|$ is the voltage magnitude at bus $n$ (p.u.), and $\theta_n$ is the voltage angle at bus $n$. (The p.u. is a standard normalization of various quantities in a power system.)

The DC linearization of (26) and (27) is obtained by assuming that the variations in voltage angle and magnitude between the two buses are small, and that the conductance is negligible:

$$\cos(\theta_n - \theta_m) \approx 1, \ \sin(\theta_n - \theta_m) \approx \theta_n - \theta_m,$$

$$|V_n| = |V_m| = \tilde{V} \text{ with } \tilde{V} \text{ constant, and } G_{nm} \approx 0.$$

The DC power flow equations are thus

$$P_{nm} = -\tilde{V}^2 B_{nm}(\theta_n - \theta_m),$$
(28)

$$Q_{nm} = 0.$$
(29)

While the DC linearization has many advantages in practice, it does have the drawback that it provides no information about the reactive power component of the power flow.

Koster and Lemkens [28] consider a different linearization where, unlike in the DC approach, the voltage change and the conductance are not neglected. This alternative linearization comes from two PhD theses at RWTH Aachen [11, 39]. The idea is to introduce new variables $\Delta V_{nm} = V_n - V_m$ and $\Delta\theta_{nm} = \theta_n - \theta_m$ and to maintain the assumptions of the DC linearization except for $\Delta V_{nm} \approx 0$ and $\Delta\theta_{nm} \approx 0$. The result is the pair of equations

$$P_{nm} = \tilde{V}G_{nm}\Delta V_{nm} - \tilde{V}^2 B_{nm}\Delta\theta_{nm},$$

$$Q_{nm} = -\tilde{V}B_{nm}\Delta V_{nm} - \tilde{V}^2 G_{nm}\Delta\theta_{nm}.$$

Compared to the DC equations, this linearization has the advantage that it provides a value for the reactive power flow. The linearization is applied to a power network design problem in [28] and the quality of the solutions obtained is superior to those from the DC linearization.

Taylor and Hover [53] studied an RLT-based linearization [50] of the AC equations in the context of transmission system planning. The idea here is to reformulate the power flow equations in polynomial form using the rectangular coordinates for the voltage variables. If we express the voltage at bus $n$ in terms of real and imaginary components, i.e., $V_n = e_n + j f_n$, where $j = \sqrt{-1}$, then $|V_n| = e_n^2 + f_n^2$, $\theta_n = \arctan \frac{f_n}{e_n}$, and we can rewrite equations (26) and (27) in the form

$$P_{nm} = \frac{1}{2} G_{nm}(e_n^2 + f_n^2 - e_m^2 - f_m^2) + B_{nm}(e_n f_m - e_m e_n),$$

$$Q_{nm} = -\frac{1}{2}(B_{nm} + B_{nm}^{sh})(e_n^2 + f_n^2 - e_m^2 - f_m^2) + G_{nm}(e_n f_m - e_m e_n),$$

where $B_{nm}^{sh}$ is the shunt susceptance of the line. Using this formulation of the power flow equations in a model for the planning problem, the results in [53] show improvements over existing DC-based models. However, the approach suffers from the rapid growth in the numbers of variables and of constraints.

An alternative way to construct linearizations of general polynomial optimization problems is via semidefinite optimization. Lavaei and Low [29] consider an exact formulation of the optimal power flow problem using a semidefinite approach and a rank-one (or equivalent) constraint on the matrix variable that guarantees exactness. By omitting this rank constraint, they obtain a semidefinite relaxation and hence the corresponding dual relaxation. They prove that if the duality gap is zero, then the dual optimal solution provides an optimal solution for the power flow problem. While it is observed in [29] that this condition on the duality gap holds for several IEEE benchmark systems, Lesieutre et al. [31] show that it can fail for some cases of practical interest. Nevertheless the results in [31] suggest that the semidefinite approach has the potential to assist in efficiently identifying a variety of solutions to the power flow equations.

Independently from the work in [29, 31], there has been tremendous progress recently on the theory of semidefinite relaxations, and conic relaxations more generally, for polynomial optimization problems [1]. It would thus be interesting to investigate how to apply the conic/semidefinite approach in the UC context. One challenging issue is that the semidefinite approaches also incur a rapid growth in the size of the matrix variables and the number of constraints.

Most recently, Coffrin and Van Hentenryck [13] present a novel approach to approximate the AC power flow equations. Specifically, they present three different linear approximations of the AC power flow equations. These approximations are based on:

- accounting for the voltage phase angles and the voltage magnitudes (coupled via the equations for active and reactive power)and
- using a piecewise linear approximation of the cosine term in the power flow equations

The three approximations are referred to as *hot-start*, *warm-start*, and *cold-start*. They differ depending on the voltage information available before the power flow is calculated. The hot-start approximation assumes that a base-point AC solution is known, and hence the model can use additional information such as voltage magnitudes. On the other hand, the cold-start approximation assumes no voltage information is available, while the warm-start approximation assumes that target voltages are given (though an actual solution may not exist for the target values).

We briefly outline the hot-start approximation. We assume that we are given voltage magnitudes $\tilde{V}_n$ at each bus $n$. The idea is to approximate $\sin(x)$ by $x$ as previously and to better approximate cos by using a piecewise linear function $\widehat{\cos}$ (on a suitable domain). Substituting these ingredients into (26) and (27) we obtain the equations of the hot-start linear approximation of the AC power flow:

$$P_{nm} = \tilde{V}_n^2 G_{nm} - \tilde{V}_n \tilde{V}_m G_{nm} \widehat{\cos}(\theta_n - \theta_m) - \tilde{V}_n \tilde{V}_m B_{nm}(\theta_n - \theta_m), \qquad (30)$$

$$Q_{nm} = -\tilde{V}_n^2 B_{nm} + \tilde{V}_n \tilde{V}_m B_{nm} \widehat{\cos}(\theta_n - \theta_m) - \tilde{V}_n \tilde{V}_m G_{nm}(\theta_n - \theta_m). \qquad (31)$$

It remains to specify how to build the piecewise linear approximation $\widehat{\cos}$. Coffrin and Van Hentenryck use a small number of tangent planes to $\cos(x)$ on the domain $(-\frac{\pi}{2}, \frac{\pi}{2})$ or on a subset of it if possible, as this will tighten the approximation. The results reported in [13] suggest that the three approximations proposed improve the accuracy of power flow calculations for power restoration and capacitor placement problems.

It would be interesting to incorporate one of the AC power flow approximations into the basic UC model and compare the results with those obtained via other techniques such as the Benders-based approach or, for small- and medium-sized instances, with a semidefinite-based approach.

## 5   Demand Uncertainty and Intermittent Generation

In this last section we briefly address developments concerned with handling uncertainty in the UC context. There are fundamentally two sources of uncertainty when making UC decisions, namely the demand forecast and generator failure. We will restrict our attention here to the uncertainty arising from demand forecasting errors. This aspect is currently of great interest because the uncertainty in (net) demand includes the variability in generation from intermittent sources such as wind and solar.

Demand forecast fluctuations have traditionally been handled via the level of reserve generation required (see, e.g., [9]). While this approach is straightforward to implement in practice, it does not explicitly account for the uncertainty. It can also be economically inefficient, and while these inefficiencies may have been acceptable in the past, the growing provision of power from intermittent generation increases the cost of this inefficiency and raises the importance of improved means to model the uncertainty.

With multiple jurisdictions around the world experiencing, and in some cases mandating, significant increases in the proportion of electricity generated by intermittent sources, the importance of models incorporating uncertainty is increasing.

For example, one such model was recently used to predict the operational impacts of high levels of wind power generation in Ireland [37]. As there is a large body of relevant research even if we restrict our attention to UC, we choose to only summarize three techniques that have been proposed to handle this uncertainty: stochastic optimization, chance-constrained optimization (CCO), and robust optimization. Our objective is to give the interested reader an entry point into the area.

## 5.1   Stochastic Optimization

A well-known optimization technique for managing uncertainty is stochastic optimization (SO). A basic description of the SO approach is that the uncertainty is represented by a set of scenarios that describe possible future evolutions of demand over the planning period; each scenario is assigned a probability; and the optimization objective is to minimize the expected production cost over the set of scenarios. More specifically, if we let $S$ denote the set of scenarios, $\sigma \in S$ denote one of those scenarios, $\xi_\sigma$ denote the probability of scenario $\sigma$, and $D(\sigma, t)$ and $R(\sigma, t)$ denote the demand and spinning reserve at time period $t$ under scenario $\sigma$, then the variables are instanciated for each scenario:

$$p_g(\sigma, t) = \text{power produced by generator } g \text{ at time } t \text{ under scenario } \sigma,$$

$$s_g(\sigma, t) = \begin{cases} 1, & \text{if generator } g \text{ is on at time period } t \text{ under scenario } \sigma, \\ 0, & \text{otherwise}, \end{cases}$$

$$s_g^{su}(\sigma, t) = \begin{cases} 1, & \text{if generator } g \text{ starts up at time period } t \text{ under scenario } \sigma, \\ 0, & \text{otherwise}, \end{cases}$$

$$s_g^{sd}(\sigma, t) = \begin{cases} 1, & \text{if generator } g \text{ shuts down at time period } t \text{ under scenario } \sigma, \\ 0, & \text{otherwise}, \end{cases}$$

Furthermore the objective function (3) is rewritten as

$$\sum_{\sigma \in S} \xi_\sigma \left( \sum_{t \in T} \sum_{g \in G} c_g^f(p_g(\sigma, t)) \, s_g(\sigma, t) + c_g^{su} s_g^{su}(\sigma, t) + c_g^{sd} s_g^{sd}(\sigma, t) \right) \qquad (32)$$

and the constraints are adjusted accordingly, for example, constraint (4) becomes

$$\sum_{g \in G} p_g(\sigma, t) \geq D(\sigma, t) + R(\sigma, t), \forall t \in T, \forall \sigma \in S. \qquad (33)$$

While most SO models do not include reserve requirements, an SO approach taking reserves into account was proposed in [47].

The SO models can incorporate uncertainties and risks of various kinds. For example, in the UC context, SO models have been used to model the impacts of high levels of wind penetration [37, 55] and the cost of ensuring system reliability [60]. Another advantage of SO is that even though UC decisions are made well ahead of real-time operation, the SO approach can model the anticipated reactions of the system operator when more information becomes available in real time. This was already observed when the first SO formulation for UC was proposed in [57] and is done using a two-stage SO approach. The application of two-stage SO in UC is based on the idea that in the first stage UC decisions are made using only the demand information available hours or days in advance of real-time operations, and in the second stage the uncertainty is realized and an optimal economic dispatch (and its corresponding cost) can be computed. Since the second-stage information is scenario-dependent, the optimization model minimizes the deterministic cost of the first-stage decisions (including, e.g., the cost of the spinning reserves scheduled) plus the expected cost of the second-stage decisions (including the cost of the reserves actually called upon according to each scenario).

The quality of the solutions obtained using SO critically depends on the choice of $S$ in the sense that having more scenarios usually leads to a more accurate model. However, a larger set of scenarios also means a greater number of variables and constraints. The growth in the size and number of MILP problems that must be solved in a two-stage SO is one of its major drawbacks because it means that SO approaches are computationally expensive. Another drawback of the SO approach is that it assumes explicit knowledge of the probability distribution of the (uncertain) demand. In practice this distribution is estimated empirically based on past data and experience and/or using simulation models (see, e.g., [59]). The limitations of the probability estimation may impact the quality of the results delivered by the SO model.

## 5.2 Chance-Constrained Optimization

The principle of CCO is to require that one or more constraints be satisfied with a given probability. The use of CCO in the context of UC was proposed in [44] to address uncertainty in the demand. Supposing that demand is uncertain, we view it as a random variable and replace constraint (4) by

$$\Pr\left\{\sum_{g \in G} p_g(t) \geq D(t) + R(t), \forall t \in T\right\} \geq 1 - \alpha, \tag{34}$$

where $\alpha$ is the (given) probability threshold with which the constraint may fail to hold.

Because the constraint (34) cannot be implemented as stated, different ways have been proposed to handle it in practice [36, 44, 56]. Let us illustrate one way to convert the constraint (34) into an equivalent deterministic form. If we ignore the reserves $R(t)$ and assume that demand follows a normal distribution then we can use the cumulative distribution function $\Phi$ of the normal distribution to compute $Z_\alpha$ such that $\Phi(Z_\alpha) = 1 - \alpha$, then the deterministic equivalent of (34) is

$$\sum_{g \in G} p_g(t) \geq Z_\alpha, \forall t \in T.$$

The CCO can in principle provide solutions that are more robust than those obtained by SO because there is no more dependency on the choice of scenarios and hence on the potential impact of unexpected events. On the other hand, the CCO suffers from most of the same limitations as SO; in particular, a CCO model for demand uncertainty requires explicit knowledge of the probability distribution of the demand. Furthermore, CCO models are often difficult to solve, while SO are straightforward to solve for each given scenario.

## 5.3   Robust Optimization

Robust optimization (RO) is an alternative technique that aims to overcome some of the aforementioned drawbacks of SO and CCO. The idea behind RO is to avoid the assumption of explicit knowledge of the probability distribution by using a *deterministic* uncertainty set. This uncertainty set can be defined using only limited information about the uncertain quantity, namely the mean together with some estimate of the variance or a range of possible variation around the mean. (If additional information is available, it can usually be incorporated into a robust framework and hence improve the quality of the model. The point is that additional information is not necessary to apply RO.) Once the uncertainty set is chosen, the RO model computes an optimal solution that protects the operator against every possible realization of the demand contained in the chosen set. In this sense, RO is more conservative than SO, and thus it is consistent with the accepted practice in power system operations.

We first sketch the main ideas for the application of RO to UC as proposed in [7]. For ease of notation, let $D = (D(t))_t$ be the vector of demands for the planning period. Similarly define $\bar{D} = (\bar{D}(t))_t$ as the vector of expected demands and $\hat{D} = (\hat{D}(t))_t$ as the vector of maximum allowed deviation ($+$ or $-$) from the expected demand at time period $t$. The simplest uncertainty set is the hypercube defined by the full range of possible values:

$$\mathcal{H}(\bar{D}, \hat{D}) = \left\{ D \in \mathbb{R}^T \mid D(t) \in [\bar{D}(t) - \hat{D}(t), \bar{D}(t) + \hat{D}(t)] \ \forall t \in T \right\}. \quad (35)$$

This type of uncertainty set is usually too pessimistic. To achieve a better trade-off in the optimization, one may introduce the concept of a *budget of uncertainty*. This is a quantity that bounds the total amount by which the solution may deviate from the expected demand. Let $\mu$ denote this quantity, then the uncertainty set is

$$\mathcal{D}(\bar{D}, \hat{D}) = \mathcal{H}(\bar{D}, \hat{D}) \cap \left\{ D \in \mathbb{R}^T \mid \sum_{t \in T} \left( \frac{|D(t) - \bar{D}(t)|}{\hat{D}(t)} \right) \leq \mu \right\}. \quad (36)$$

The budget of uncertainty has an elegant practical interpretation. When $\mu = 0$ the uncertainty set allows only $D(t) = \bar{D}(t)$ for all $t \in T$, i.e., this is the deterministic case. As $\mu$ increases, the set $\mathcal{D}(\bar{D}, \hat{D})$ increases, and when $\mu = T$, $\mathcal{D}(\bar{D}, \hat{D})$ is the entire hypercube of possible values.

The RO approach was used in various papers, e.g. [23, 62, 63], to model uncertainty in wind power generation. Although similar in spirit, the uncertainty set used in [23,62] has important differences from the set in [7]. Specifically, the construction in [23,62] assumes that the wind power output lies in the interval

$$[\bar{W}_n(t) - \hat{W}_n^-(t), \bar{W}_n(t) + \hat{W}_n^+(t)],$$

where $\bar{W}_n(t)$ is the forecasted wind power output at bus $n$ for the time period $t$ and $\hat{W}_n^-(t)$ and $\hat{W}n^+(t)$ are the maximum allowed deviations from the forecast. Note that in contrast with the intervals in (35), the intervals here may not be centered at the expected value. Furthermore, the budget of uncertainty is defined as a *wind cardinality budget*, meaning that the quantity budgeted is not the amount by which the wind power generation deviates from the forecast but rather the number of time periods at which this happens at each bus. Specifically, the wind power output $W^n(t)$ at bus $n$ and time period $t$ is expressed as

$$W^n(t) = \bar{W}_n(t) + z_n^+(t)\hat{W}_n^+(t) - z_n^-(t)\hat{W}_n^-(t),$$

where the variables $z_n^+(t)$ and $z_n^-(t)$ are binary, and the wind cardinality budget takes the form

$$\sum_{t \in T}(z_n^+(t) + z_n^-(t)) \leq \mu_n,$$

where $\mu_n$ is the wind cardinality budget for bus $n$.

## 5.4 Combining Techniques

We conclude by pointing out that different techniques can be combined to increase the quality of the model. For instance, SO and CCO are successfully used together in [56]. However this increases the complexity of the models and hence the computational effort required to solve them. As the importance of modeling uncertainty in UC is undeniable and research will continue for the foreseeable future, the design of practical models for UC incorporating uncertainty remains a challenge for optimization researchers that is of great importance for power engineers.

# 6  Conclusion

More than 50 years have passed since MILP was first proposed as a means to inform UC decisions [4]. There have been tremendous improvements in MILP technology since then, and MILP has now become a standard tool for power system operators. Nevertheless the basic UC problem still poses challenges for researchers. This article summarized several modeling developments that have the potential to significantly contribute to addressing some of these challenges.

There are other challenges that were not covered here but are nevertheless of significant practical importance. One of them is that there is a real need to solve day-ahead planning problems on the basis of 15-min intervals rather than 1-h. While this is currently beyond the capabilities of state-of-the-art integer optimization, its implementation may be helped by the fact that in practice it is unlikely that all the units would be evaluated at every step, and therefore working with a subset of the generators available might suffice. Another is that it is imperative to successfully handle the model complexity arising from real-world needs such as the co-optimization of UC and of the system topology (see, e.g., [20, 27]).

In conclusion, there remain plenty of opportunities for optimization researchers to make major contributions to modeling and solving this challenging class of power engineering problems.

# References

1. Anjos, M.F., Lasserre, J.B. (eds.): Handbook on semidefinite, conic and polynomial optimization. In: International Series in Operations Research and Management Science. Springer, New York (2012)
2. Arroyo, J.M., Conejo, A.J.: Optimal response of a thermal unit to an electricity spot market. IEEE Trans. Power Syst. 15(3), 1098–1104 (2000)
3. Arroyo, J.M., Conejo, A.J.: Modeling of start-up and shut-down power trajectories of thermal units. IEEE Trans. Power Syst. 19(3), 1562–1568 (2004)
4. Baldwin, C.J., Dale, K.M., Dittrich, R.F.: A study of the economic shutdown of generating units in daily dispatch. Power apparatus and systems, part III. Trans. Am. Inst. Electrical Engineers 78(4), 1272–1282 (1959)
5. Bautista, G., Anjos, M.F., Vannelli, A.: Formulation of oligopolistic competition in AC power networks: an NLP approach. IEEE Trans. Power Syst. 22(1), 105–115 (2007)
6. Bautista, G., Anjos, M.F., Vannelli, A.: Modeling market power in electricity markets: is the devil only in the details? Electricity J 20(1), 82–92 (2007)
7. Bertsimas, D., Litvinov, E., Sun, X.A., Zhao, J., Zheng, T.: Adaptive robust optimization for the security constrained unit commitment problem. IEEE Trans. Power Syst. 28(1), 52–63 (2013)

8. Bhattacharya, K., Bollen, M.H.J., Daalder, J.E.: Operation of Restructured Power Systems. Springer, New York (2001)
9. Billinton, R., Allan, R.N.: Reliability Evaluation of Power Systems, 2nd edn. Plenum Press, New York (1984)
10. Borghetti, A., Frangioni, A., Lacalandra, F., Nucci, C.A., Pelacchi, P.: Using of a cost-based unit commitment algorithm to assist bidding strategy decisions. In: Power Tech Conference Proceedings, vol. 2 (2003)
11. Braun, A.: Anlagen- und Strukturoptimierung von 110-kV-Netzen. PhD thesis, RWTH Aachen University (2001)
12. Carrion, M., Arroyo, J.M.: A computationally efficient mixed-integer linear formulation for the thermal unit commitment problem. IEEE Trans. Power Syst. **21**(3), 1371–1378 (2006)
13. Coffrin, C., Van Hentenryck, P.: A linear-programming approximation of ac power flows. Technical report, CoRR, abs/1206.3614 (2012)
14. Costanzo, G.T., Zhu, G., Anjos, M.F., Savard, G.: A system architecture for autonomous demand side load management in smart buildings. IEEE Trans. Smart Grid **3**(4), 2157–2165 (2012)
15. Dennis, Jr, J.E., Schnabel, R.B.: Numerical Methods for Unconstrained Optimization and Nonlinear Equations. Number 16 in Classics in Applied Mathematics. SIAM (1996)
16. Frangioni, A., Gentile, C.: Perspective cuts for a class of convex 0-1 mixed integer programs. Math. Program. **106**(2), 225–236 (2006)
17. Frangioni, A., Gentile, C., Lacalandra, F.: Tighter approximated MILP formulations for unit commitment problems. IEEE Trans. Power Syst. **24**(1), 105–113 (2009)
18. Fu, Y., Shahidehpour, M., Li, Z.: Security-constrained unit commitment with AC constraints*. IEEE Trans. Power Syst. **20**(3), 1538–1550 (2005) [*Corrected version of **20**(2), 1001–1013]
19. Fu, Y., Shahidehpour, M., Li, Z.: AC contingency dispatch based on security-constrained unit commitment. IEEE Trans. Power Syst. **21**(2), 897–908 (2006)
20. Hedman, K.W., Ferris, M.C., O'Neill, R.P., Fisher, E.B., Oren, S.S.: Co-optimization of generation unit commitment and transmission switching with N-1 reliability. IEEE Trans. Power Syst. **25**(2), 1052–1063 (2010)
21. Hobbs, B.F., Rothkopf, M.H., O'Neill, R.P., Chao, H.-P. (eds.): The next generation of electric power unit commitment models. In: International Series in Operations Research & Management Science. Kluwer Academic Publishers, Norwell (2001)
22. Jeroslow, R.: Trivial integer programs unsolvable by branch-and-bound. Math. Program. **6**, 105–109 (1974)
23. Jiang, R., Wang, J., Guan, Y.: Robust unit commitment with wind power and pumped storage hydro. IEEE Trans. Power Syst. **27**(2), 800–810 (2012)
24. Kaibel, V., Loos, A.: Branched polyhedral systems. In: IPCO 2010: The Fourteenth Conference on Integer Programming and Combinatorial Optimization, vol. 6080 of Lecture Notes in Computer Science, pp. 177–190. Springer (2010)
25. Kaibel, V., Pfetsch, M.E.: Packing and partitioning orbitopes. Math. Program. **114**, 1–36 (2008)
26. Kaibel, V., Peinhardt, M., Pfetsch, M.E.: Orbitopal fixing. Discrete Optim. **8**(4), 595–610 (2011)
27. Khodaei, A., Shahidehpour, M.: Transmission switching in security-constrained unit commitment. IEEE Trans. Power Syst. **25**(4), 1937–1945 (2010)
28. Koster, A.M.C.A., Lemkens, S.: Designing AC power grids using integer linear programming. In: Pahl, J., Reiners, T., Voß, S. (eds.) Network Optimization, vol. 6701 of Lecture Notes in Computer Science, pp. 478–483 (2011)
29. Lavaei, J., Low, S.H.: Zero duality gap in optimal power flow problem. IEEE Trans. Power Syst. **27**(1), 92–107 (2012)
30. Lee, J., Leung, J., Margot, F.: Min-up/min-down polytopes. Discrete Optim. **1**(1), 77–85 (2004)

31. Lesieutre, B.C., Molzahn, D.K., Borden, A.R., DeMarco, C.L.: Examining the limits of the application of semidefinite programming to power flow problems. In: Proceedings of Communication, Control, and Computing (Allerton), pp. 1492–1499 (2011)
32. Liu, C., Shahidehpour, M., Li, Z., Fotuhi-Firuzabad, M.: Component and mode models for the short-term scheduling of combined-cycle units. IEEE Trans. Power Syst. **24**(2), 976–990 (2009)
33. Lotfjou, A., Shahidehpour, M., Fu, Y., Li, Z.: Security-constrained unit commitment with AC/DC transmission systems. IEEE Trans. Power Syst. **25**(1) 531–542 (2010)
34. Margot, F.: Pruning by isomorphism in branch-and-cut. Math. Program. **94**, 71–90 (2002)
35. Margot, F.: Exploiting orbits in symmetric ILP. Math. Program. Series B **98**, 3–21 (2003)
36. Mazadi, M., Rosehart, W.D., Malik, O.P., Aguado, J.A.: Modified chance-constrained optimization applied to the generation expansion problem. IEEE Trans. Power Syst. **24**(3) 1635–1636 (2009)
37. Meibom, P., Barth, R., Hasche, B., Brand, H., Weber, C., O'Malley, M.: Stochastic optimization model to study the operational impacts of high wind penetrations in ireland. IEEE Trans. Power Syst. **26**(3), 1367–1379 (2011)
38. Morales-España, G., Latorre, J.M., Ramos, A.: Tight and compact MILP formulation of start-up and shut-down ramping in unit commitment. IEEE Trans. Power Syst. **28**(2), 1288–1296 (2013)
39. Moser, A.: Langfristig optimale Struktur und Betriebsmittelwahl für 110-kV-Überlandnetze. PhD thesis, RWTH Aachen University (1995)
40. Ostrowski, J., Anjos, M.F., Vannelli, A.: Modified orbital branching with applications to orbitopes and to unit commitment. Cahier du GERAD G-2012-61, GERAD, Montreal, QC, Canada (2012)
41. Ostrowski, J., Anjos, M.F., Vannelli, A.: Tight mixed integer linear programming formulations for the unit commitment problem. IEEE Trans. Power Syst. **27**(1), 39–46 (2012)
42. Ostrowski, J., Linderoth, J., Rossi, F., Smriglio, S.: Orbital branching. Math. Program. **126**(1), 147–178 (2009)
43. Ott, A.L.: Evolution of computing requirements in the PJM market: past and future. In: Power and Energy Society General Meeting, 2010 IEEE, pp. 1 –4 (2010)
44. Ozturk, U.A., Mazumdar, M., Norman, B.A.: A solution to the stochastic unit commitment problem using chance constrained programming. IEEE Trans. Power Syst. **19**(3), 1589–1598 (2004)
45. Padhy, N.P.: Unit commitment-a bibliographical survey. IEEE Trans. Power Syst. **19**(2), 1196–1205 (2004)
46. Rajan, D., Takriti, S.: Minimum up/down polytopes of the unit commitment problem with start-up costs. Technical report, IBM Research Report (2005)
47. Ruiz, P.A., Philbrick, C.R., Zak, E., Cheung, K.W., Sauer, P.W.: Uncertainty management in the unit commitment problem. IEEE Trans. Power Syst. **24**(2), 642–651 (2009)
48. Shahidehpour, M., Yamin, H., Li, Z.: Market operations in electric power systems: forecasting, scheduling, and risk management. Wiley, New York (2003)
49. Sherali, H.D., Smith, J.C.: Improving zero-one model representations via symmetry consider-ations. Manag. Sci. **47**(10), 1396–1407 (2001)
50. Sherali, H.D., Tuncbilek, C.H.: A global optimization algorithm for polynomial programming problems using a reformulation-linearization technique. J. Global Optim. **2**, 101–112 (1992)
51. Simoglou, C.K., Biskas, P.N., Bakirtzis, A.G.: Optimal self-scheduling of a thermal producer in short-term electricity markets by MILP. IEEE Trans. Power Syst. **25**(4), 1965–1977 (2010)
52. Sioshansi, R., O'Neill, R., Oren, S.S.: Economic consequences of alternative solution methods for centralized unit commitment in day-ahead electricity markets. IEEE Trans. Power Syst. **23**(2), 344–352 (2008)
53. Taylor, J.A., Hover, F.S.: Linear relaxations for transmission system planning. IEEE Trans. Power Syst. **26**(4), 2533–2538 (2011)
54. Troy, N., Denny, E., O'Malley, M.: Base-load cycling on a system with significant wind penetration. IEEE Trans. Power Syst. **25**(2), 1088–1097 (2010)

55. Tuohy, A., Meibom, P., Denny, E., O'Malley, M.: Unit commitment for systems with significant wind penetration. IEEE Trans. Power Syst. **24**(2), 592–601 (2009)
56. Wang, Q., Guan, Y., Wang, J.: A chance-constrained two-stage stochastic program for unit commitment with uncertain wind power output. IEEE Trans. Power Syst. **27**(1), 206–215 (2012)
57. Wiebking, R.: Stochastische modelle zur optimalen lastverteilung in einem kraftwerksverbund. Zeitschrift für Oper. Res. **21**(6), B197–B217 (1977)
58. Wu, L.: A tighter piecewise linear approximation of quadratic cost curves for unit commitment problems. IEEE Trans. Power Syst. **26**(4), 2581–2583 (2011)
59. Wu, L., Shahidehpour, M., Tao, L.: Stochastic security-constrained unit commitment. IEEE Trans. Power Syst. **22**(2), 800–811 (2007)
60. Wu, L., Shahidehpour, M., Li, T.: Cost of reliability analysis based on stochastic unit commitment. IEEE Trans. Power Syst. **23**(3), 1364–1374 (2008)
61. Yamin, H.Y.: Review on methods of generation scheduling in electric power systems. Electric Power Syst. Res. **69**(2–3), 227–248 (2004)
62. Zhao, C., Wang, J., Watson, J.-P., Guan, Y.: Multi-stage robust unit commitment considering wind and demand response uncertainties. IEEE Trans. Power Syst. **28**(3), 2708–2717 (2013)
63. Zhao, L., Zeng, B.: Robust unit commitment problem with demand response and wind energy. In: Power and Energy Society General Meeting, 2012 IEEE, pp. 1–8 (2012)

# Portfolio Optimization with Combinatorial and Downside Return Constraints

Miguel A. Lejeune

**Abstract** We study a probabilistic portfolio optimization model in which trading restrictions modeled with combinatorial constraints are accounted for. We provide several deterministic reformulations equivalent to this stochastic programming problem and discuss their computational efficiency. The reformulated problem takes the form of a mixed-integer nonlinear problem and is solved with an exact outer approximation algorithm. This latter is based on the early recognition of the problem structure and permits a hierarchical organization of the computations. Computational tests show the contribution of the proposed algorithm that outperforms the Cplex 12.4 solver in terms of computational time and quality of the obtained solutions.

**Keywords** Stochastic portfolio optimization • Probabilistic Markowitz • Combinatorial trading constraints • Stochastic programming • Outer approximation algorithm

## 1 Introduction

Much effort has been devoted in the last decade to the extension of the mean–variance portfolio optimization model proposed by Markowitz [17] in the late 1950s (see [9] for a recent review). The mean–variance approach trades off expected returns against variance with this latter used as the risk measure reflecting the volatility of the market.

Consider $n$ assets with position given by the vector $w$. Let $\mu \in \mathcal{R}^n$ be the $n$-dimensional mean return. The Markowitz model assumes that the expected returns

M.A. Lejeune (✉)
Department of Decision Sciences, George Washington University,
2201 G Street, Suite 415, NW Washington, DC, USA
e-mail: mlejeune@gwu.edu

L.F. Zuluaga and T. Terlaky (eds.), *Modeling and Optimization: Theory and Applications*,
Springer Proceedings in Mathematics & Statistics 62, DOI 10.1007/978-1-4614-8987-0_2,
© Springer Science+Business Media New York 2013

and the components of the positive definite variance–covariance return matrix $\Sigma \in \mathcal{R}^{n \times n}$ are known. Each of the three variants of the mean–variance model is formulated as a convex nonlinear optimization problem. One of them involves the construction of a portfolio with minimal portfolio variance $w^T \Sigma w$ provided that a prescribed return level $R$ is attained and can be formulated as the following mathematical program:

$$\textbf{MV}: \quad \min \quad w^T \Sigma w \tag{1}$$

$$\text{subject to} \quad \mu^T w \geq R \tag{2}$$

$$w \in \mathcal{X}. \tag{3}$$

The notation $\mathcal{X}$ refers to the polytope defining the set of admissible portfolios

$$\mathcal{X} = \left\{ w \in \mathbb{R}^n : e^T w = 1, \ w \geq 0 \right\}, \tag{4}$$

and $e$ is an all-one vector or appropriate dimension. Optimal portfolios located on the efficient frontier are those exposing the investor to the minimum possible risk (i.e., variance) and providing a specified return level $R$ in (2). The first constraint in (4) ensures that the totality of the capital is invested in the $n$ assets while the other constraints prevents short selling.

An issue commonly associated with the Markowitz model is the estimation risk which refers to the sensitivity of the mean–variance optimal portfolios with respect to the single-point estimate of the parameters (expected returns and variance–covariance matrix) of the model. Multiple empirical studies have shown that minor perturbations in the estimates of these parameters can lead to a drastically different optimal portfolio. In practice, investors prefer to trade off some return for more safety and construct a portfolio that performs well under diverse market conditions. Models based on stochastic dominance [5] and programming [21], robust optimization [6], or robust statistics [22] have been designed in order to obtain allocation policies that are less affected by the side effects of the estimation risk. In this study, we focus on the probabilistic asset allocation model **PMV** introduced by Bonami and Lejeune [2] that takes into account the difficulty to estimate asset returns and the risk to rely on a one-point estimate. Accordingly, the **PMV** model accounts for the incomplete knowledge of the return behavior and defines the vector of asset returns as a vector $\xi$ of random variables. The model **PMV** takes the form of a stochastic programming problem with random technology matrix in which the decision variables $w$ are multiplied by stochastic coefficients $\xi$ that are not (necessarily) independent

$$\textbf{PMV}: \quad \min \quad w^T \Sigma w$$

$$\text{subject to} \quad \mathbb{P}(\xi^T w \geq R) \geq p \tag{5}$$

$$w \in \mathcal{X}.$$

The symbol $\mathbb{P}$ refers to a probability measure while $p$ is the specified probability level typically defined on $[0.7, 1)$. The asset allocation model **PMV** implements a downside risk measure that probabilistically prevents the return of the portfolio to fall below the return level $R$ in (5) and can be viewed as a probabilistic version of the Markowitz model. The risk measure is closely related to Roy's safety-first risk metric [20] and to Kataoka's model [11].

Besides the estimation risk concern, the Markowitz model is often amended by portfolio managers to integrate specific institutional features and trading criteria [6]. We describe in this paper some of the most common practical investment requirements that are formulated with combinatorial constraints. Those include the presence of nonproportional transaction costs, the requirements to invest a minimal amount in any selected asset (buy-in threshold constraint), to invest in a minimum number of asset classes or industrial sectors (diversification constraint), to restrict the number of positions hold (cardinality constraint), and to buy shares by lots of batches (round lot constraint). As we shall see in Sect. 3, these practical considerations require the use of integer decision variables and the introduction of combinatorial constraints in model **PMV**, transforming it into a stochastic integer problem and further compounding its numerical solution.

Our contributions are twofold. On the modeling side, we provide a series of deterministic and convex formulations equivalent to the probabilistic portfolio optimization problem and analyze their features (Sect. 2). The analysis is based on earlier results presented in [2, 7]. We also review some of the most common trading restrictions and describe how they can be formulated with combinatorial constraints (Sect. 3). On the algorithmic side, we develop a variant of the exact outer approximation algorithm proposed in [13] (see Sect. 4). Computational tests indicate the computational benefits of using the proposed algorithm (Sect. 5).

## 2 Model Reformulations

Under the formulation **PMV**, the probabilistic portfolio optimization problem cannot be handled by any optimization solvers. We shall now review a number of conic reformulations (see [2, 7]) that are greatly beneficial for the numerical solution of problem **PMV**.

### 2.1 Nonlinear Reformulation

Let $\psi = \frac{\xi^T w - \mu^T w}{\sqrt{w^T \Sigma w}}$ be a random variable with mean 0 and variance 1 representing the normalized portfolio return. Further, we denote by $F_{(w)}$ the cumulative probability distribution of $\psi$, by $F_{(w)}^{-1}$ its inverse, and by $F_{(w)}^{-1}(1-p)$ its $(1-p)$−quantile. The subscript $w$ indicates that the exact form of the probability distribution $F$ depends on the holdings $w$ of the portfolio.

**Theorem 1 (Kataoka [11]).** *The nonlinear optimization problem* **C1**

$$\textbf{C1}: \quad \min \quad w^T \Sigma w$$

$$\text{subject to } \mu^T w + F_{(w)}^{-1}(1 - p)\sqrt{w^T \Sigma w} \geq R \tag{6}$$

$$w \in \mathcal{X}$$

*is equivalent to* **PMV**.

The next subsection is based on Sect. 2.2.1 in [2] and studies under which conditions the feasible set defined by the nonlinear constraint (6) is convex, thereby making **C1** a convex optimization problem.

## 2.2  Convexity

*(a) Symmetric Probability Distributions*
Theorem 2 studies the convexity of the feasible set defined by (6) for symmetric probability distributions.

**Theorem 2 (Bonami and Lejeune [2]).** *If $p \in [0.5, 1)$ and if the probability distribution of $\xi^T w$ is symmetric, the constraint $\mu^T w + F_{(w)}^{-1}(1 - p)\sqrt{w^T \Sigma w} \geq R$ equivalent to (5) is a second-order cone constraint.*

Thus, problem **C1** minimizes a convex quadratic function over a second-order cone and some linear constraints and is therefore a convex problem.

*(b) Positively Skewed Probability Distributions*
Skewness is typically used to measure the asymmetry of a probability distribution. A probability distribution is said to be right-skewed or to have positive skewness (resp., left-skewed or negative skewness) if the right, upper value (resp., left, lower value) tail is longer or fatter than the left, lower value (resp., right, upper value).

**Definition 1.** Let $m$ be the median of the probability distribution $F$ of an $r$-variate random vector $\psi$ with $E[\psi] = 0$. The distribution $F$ has positive skewness if

$$\mathbb{P}(0 \geq \psi) \geq \mathbb{P}(m \geq \psi) \Leftrightarrow F^{-1}(\alpha) \leq 0, \ \alpha \leq 0.5 .$$

**Theorem 3 (Bonami and Lejeune [2]).** *If $p \in [0.5, 1)$ and if the probability distribution of $\xi^T w$ has positive skewness, then $\mu^T w + F_{(w)}^{-1}(1 - p)\sqrt{w^T \Sigma w} \geq R$ equivalent to (5) is a second-order cone constraint.*

## 2.3  Compact Reformulations of Variance

This section is based on the results proposed by Filomena and Lejeune [7] and provides two alternate ways to calculate the variance of the portfolio. To ease the notation, we shall thereafter refer to $F_{(w)}^{-1}(1 - p) = -\kappa$.

### 2.3.1  Convex Reformulations with Cholesky Decomposition

We shall first use the Cholesky decomposition to derive the convex programming problems **C2** and **E2** equivalent to **C1**. The Cholesky decomposition involves the calculation of a lower triangular matrix $C$ such that $\Sigma = CC^T$, with $\Sigma$ positive definite. Substituting $CC^T$ for $\Sigma$ in the objective function and in (6) and introducing the auxiliary nonnegative decision variable $h$ (10), we obtain the convex optimization problem **C2**:

$$\textbf{C2}: \quad \min \quad \|C^T w\|_2^2 \tag{7}$$

$$\text{subject to } (3)$$

$$\mu^T w - R \geq h \tag{8}$$

$$\kappa \, \|C^T w\|_2 \leq h, \tag{9}$$

$$h \geq 0 \tag{10}$$

where $\|x\|_2$ denotes the Euclidean norm of the vector $x$. The epigraph formulation of problem **C2** follows and has the canonical form of a second-order cone programming problem.

**Theorem 4 (Filomena and Lejeune [7]).** *Problem* **E2**

$$\textbf{E2}: \quad \min \quad h \tag{11}$$

$$\text{subject to } (3); (8); (9); (10)$$

*is equivalent to problem* **C2**.

## 2.4  Convex Reformulations with Period-Separable Formulation of Variance

The model proposed in this section uses Proposition 1 to reformulate the functions involving the variance of the portfolio as the Euclidian norm of a multidimensional vector.

**Proposition 1 (Filomena and Lejeune [7]).** *Let $M$ be the number of data points used for estimation purposes. Let $r_{jt}$ be the observed return of $j$ at $t$ and define the auxiliary variables:*

$b_t = \sum_{j=1}^{n} (r_{jt} - \mu_j) w_j$ , $t = 1, \ldots, M$. *The variance of the portfolio return becomes*

$$w^T \Sigma w = \frac{1}{M} \|b\|_2^2 .$$

The variance of the portfolio is now written in a separable form and includes $M$ squared terms $b_t$ representing each the part of the variance associated with period $t$, $t = 1, \ldots, M$.

We introduce $M$ decision variables $b_t$ unrestricted in sign and apply Proposition 1 for the reformulation of the probabilistic model as the convex optimization problem **C2**:

$$\textbf{C2}: \quad \min \quad \frac{1}{M} \|b\|_2^2 \tag{12}$$

$$\text{subject to} \quad (3); (8); (10)$$

$$\frac{\kappa}{\sqrt{M}} \|b\|_2 \leq h \tag{13}$$

$$b_t - \sum_{j=1}^{n} (r_{jt} - \mu_j) w_j = 0, \ t = 1, \ldots, M. \tag{14}$$

As for **C1**, we provide the epigraph formulation **E2** of problem **C2**:

$$\textbf{E2}: \quad \min \quad h \tag{15}$$

$$\text{subject to} \quad (3); (8); (10); (13); (14)$$

The modeling of the variance in problems **C0**, **C1**, and **E1** requires the a priori estimate of $\frac{n(n+1)}{2}$ covariance terms and can generate model specification issues, such as the obtaining of a variance–covariance matrix that is not positive semi-definite [4]. The computation of the variance proposed in this section does not have such limitations and does not make any assumption on the form or rank of $\Sigma$. Furthermore, the number of quadratic terms in problems **C2** and **E2** is much smaller than in the formulations **C0**, **C1**, and **E1** for large-scale portfolio optimization problems when the number $n$ of assets exceed by far the number $M$ of observations.

## 2.5 Equivalent Formulations and Inner Approximations Based on Probability Inequalities

In this section, we study for which classes of probability distributions the above problem is a second-order cone optimization problem (i.e., thus convex and solvable in polynomial time). Results extracted from [2, 12] are presented here and indicate that it is not always possible to derive an exact closed-form formulation of the second-order cone problem for each probability distribution. The exact value of the quantile $F_{(w)}^{-1}(1 - p)$ can be derived for some probability distributions (e.g., normal, student, uniform distribution on an ellipsoid). Often, the probability distribution of the portfolio return is only partially known and the exact value of its quantiles can only be approximated. The Cantelli [2], Chebyshev [2], and Camp–Meidell [12] probability inequalities can be used for approximating the value of the quantile, which allows for the formulation of a convex inner approximation of the probabilistic optimization problem **PMV**.

**Theorem 5 (Bonami and Lejeune [2]).** *Let the first two moments of the probability distribution of the portfolio return be finite. The second-order cone constraint*

$$\mu^T w - \sqrt{\frac{p}{1-p}} \sqrt{w^T \Sigma w} \geq R$$

*is a valid inner approximation of the probabilistic constraint* $\mathbb{P}\left(\xi^T w \geq R\right) \geq p$ *(5).*

**Theorem 6 (Lejeune [12]).** *Let the probability distribution of the portfolio return have finite first and second moments. The second-order cone constraints*

$$\mu^T w - \sqrt{\frac{1}{2(1-p)}} \sqrt{w^T \Sigma w} \geq R \tag{16}$$

*and*

$$\mu^T w - \sqrt{\frac{2}{9(1-p)}} \sqrt{w^T \Sigma w} \geq R. \tag{17}$$

*are valid inner approximations of the probabilistic constraint (5) when the probability distribution of* $\xi^T w$ *is*

- *Symmetric [for (16)]*
- *Symmetric and unimodal [for (17)]*

The constraints (16) and (17) are respectively derived from the one-sided Chebyshev inequality for symmetric probability distribution:

$$\mathbb{P}(X - \mu \geq a) \leq \frac{\sigma^2}{2a^2}, \tag{18}$$

and from the Camp–Meidell's inequality

$$P(X - \mu \geq a) \leq \frac{2\sigma^2}{9a^2}. \tag{19}$$

The expression $\min[a, b]$ refers to the minimum value of $a$ and $b$. The three inner approximations of (5) proposed above are second-order cone constraints and define a convex feasible area.

## 3  Combinatorial Constraints for Asset Allocation

Asset allocation policies are often subjected to specific constraints not present in the standard formulation of the mean–variance model. In this section, we describe some of the most common restrictions faced by asset allocation managers and modeled with combinatorial constraints (see, e.g., [2, 10, 18]). The introduction of integer variables forces to solve problem **C0**, **C1**, **E1**, **C2**, or **E2** multiple times (once at each node of a branch-and-bound tree). Filomena and Lejeune's previous computational study [7] has shown that the formulation **E2** can be solved the fastest and all subsequent formulations will be based on **E2**. We refer to [6] for a presentation of other trading restrictions (e.g., proportional transaction costs, liquidity, tracking and turnover constraints) whose consideration does not require the introduction of integer decision variables.

### 3.1  Diversification Constraints

For diversification purposes, asset managers are sometimes required to hold a representative position (i.e., at least equal to a specified level $s_{min}$) in a minimal number $l_{min}$ of asset classes, industrial sectors, and/or securities. Let $l$ be the number of industrial sectors in which positions can be hold. Every security $j$ is associated to a particular sector $k$. The sets $S_k, k = 1, \ldots, l$ of assets affiliated with a sector $k$ form a partition of $\{1, \ldots, n\}$. Let $\zeta_k$ be a binary variable taking value 1 if the portfolio includes a position in sector $k$ and taking value 0 otherwise. The feasible set defined by the diversification constraint reads

$$G_D = \left\{ (w, \zeta) \in \mathcal{R}_+^n \times \{0, 1\}^l : s_{min} \, \zeta_k \le \sum_{j \in S_k} w_j \le s_{min} + (1 - s_{min}) \, \zeta_k, k = 1, \ldots l, \ e^T \zeta \ge l_{min} \right\}.$$

(20)

The last constraint in (20) is a cardinality constraint that requires to hold positions in at least $l_{min}$ sectors.

### 3.2  Cardinality Constraints

Besides enforcing diversification goals, cardinality constraints are also needed when the asset manager is tasked (e.g., index tracking fund) to track a market benchmark and reproduce its behavior with a limited number of securities. Under such circumstances, the cardinality constraints define an upper bound $u_{max}$ on the number of positions that can be held. Let $\delta_j \in \{0, 1\}, j = 1$ denote a binary variable taking value 1 if one invests in asset $j$ and taking value 0 otherwise. The feasible set defined by such cardinality constraints is given by

$$G_C = \left\{ (w, \delta) \in \mathcal{R}_+^n \times \{0, 1\}^n : w \leq \delta \, , \, e^T \delta \leq u_{max} \right\}. \tag{21}$$

Cardinality constraints limit the number of assets included in the portfolio, thereby controlling the transaction costs. However, they are not sufficient to exclude the occurrence of small trades and are therefore often juxtaposed to buy-in threshold constraints.

## 3.3 Buy-In Threshold Constraints

Small positions have typically little impact on the performance of the portfolio, but have weak liquidity and can be costly in terms of brokerage fees, bid-ask spreads, monitoring costs, etc. Buy-in threshold constraints prevent the holding of a position if one does not invest at least $w_{min}$ which represents the minimal allowable position size. The feasible set defined by the buy-in threshold constraints reads

$$G_B = \left\{ (w, \delta) \in \mathcal{R}_+^n \times \{0, 1\}^n : w \leq \delta \, , \, w_{min} \, \delta \leq w \right\}. \tag{22}$$

## 3.4 Transaction Round Lot Constraints

Institutional investors have often to transact securities in large lots or batches of $M_j$ (e.g., 100, 500) shares. The formulation of the round lot trading requirements requires the introduction of general integer decision variables here represented by the $n$-dimensional vector $\gamma$. Let $p_j$ denote the face value of stock $j$ and $V$ be the available capital. The feasible set defined by the round lot transaction constraints is

$$G_L = \left\{ (w, \gamma) \in \mathcal{R}_+^n \times \mathcal{Z}^n : w_j = \frac{p_j \gamma_j M_j}{V} \, , \, j = 1, \ldots, n \right\}, \tag{23}$$

and imposes that the number $\gamma_j M_j$ of shares of any asset $j$ in the portfolio is a multiple of $M$. The impact of round lot transaction constraints on the structure of the portfolio can be very marked when the asset prices are large relative to the size of the trade. Transaction round lot constraints force to buy shares in large lots, thereby eliminating the risk of holding small positions.

## 3.5 Fixed Transaction Cost Constraints

Transaction costs caused, for example, by brokerage fees, liquidity costs, fund loans, and tax [15] are omitted in the standard mean–variance formulations. However, they can substantially reduce the returns of a portfolio. Accounting for them is likely to reduce the volume of trading and rebalancing operations. While sometimes modeled as proportional to the amount traded [7, 19] to represent the gap between bid and ask prices, transaction costs often involve a fixed component invariant with the transacted amount [8, 15]. We shall present a formulation of transaction costs which includes a proportional and a fixed component. We refer the reader to [16] for a comprehensive overview of functional forms used for transaction costs.

Let $w^+$ and $w^-$ denote the portfolio rebalancing quantities: $w_j^+$ is the portion of capital used to purchase $j$ and $w_j^-$ is the portion of capital obtained by selling shares of asset $j$ when constructing a new portfolio or rebalancing an existing one. We denote by $y_j$ the net position in security $j$ after rebalancing the portfolio:

$$w = w^0 + w^+ - w^- \tag{24}$$

$$w^- \le w^0 \tag{25}$$

$$w, w^+, w^- \ge 0 \tag{26}$$

The set of equalities (24) are balance constraints that ensure that the rebalanced position $w_j$ in asset $j$ is equal to the initial position $w_j^0$ increased (resp., decreased) by the purchased (resp., sold) shares $w_j^+$ (resp. $w_j^-$) of $j$. The set of constraints (25) preclude short selling when rebalancing. The proportional linear transaction cost associated to security $j$ is $c_j$ and the fixed transaction cost is $d_j$ equal to $\frac{f}{V}$, where $f$ is the fixed cost to invest in an asset and $V$ is the value of the portfolio. The probabilistic portfolio optimization model **PMV** with fixed and proportional transaction costs takes the following form:

$$\min h \tag{27}$$

$$\text{subject to } \sum_{j=1}^{n} \left( \mu_j w_j - c_j (w_j^+ + w_j^-) - d_j \delta_j \right) - R \ge h \tag{28}$$

$$\sum_{j=1}^{n} \left( w_j + c_j (w_j^+ + w_j^-) + d_j \delta_j \right) = 1 \tag{29}$$

$$w^+ + w^- \le \delta \tag{30}$$

$$\delta \in \{0, 1\}^n \tag{31}$$

$$(10); (13); (14); (24); (25); (26)$$

Constraint (28) requires the net portfolio return (after deduction of the transaction costs) to exceed the specified return level $R$. The budget constraint (29) accounts for the wealth invested in each asset and the fixed and proportional transaction costs. Constraints (30) impose each binary variable $\delta_j$ in (31) to take value 1 if the investor holds a position in the corresponding asset $j$. Note that $w^0$ is here a vector of known nonnegative parameters. If one builds a new portfolio, then each component of $w^0$ and $w^-$ are equal to 0.

Note that one can also add some buy-in threshold constraints so that the net position $w_j$ after rebalancing as well as the amounts ($w_j^+$ and $w_j^-$) traded be significant and at least equal to the minimum prescribed:

$$w \leq \delta \tag{32}$$

$$w^+ \leq \delta^+ \tag{33}$$

$$w^- \leq \delta^- \tag{34}$$

$$w_{\min} \, \delta \leq w_j \tag{35}$$

$$w_{\min} \, \delta^+ \leq w_j^+ \tag{36}$$

$$w_{\min} \, \delta^- \leq w_j^- \tag{37}$$

$$\delta^-, \delta^+ \in \{0, 1\}^n \tag{38}$$

As an alternative to transaction cost constraints, practitioners sometimes employ turnover constraints that can be linearized and can limit the turnover, and thus the transaction costs, on each individual security or over the entire portfolio.

## 4   Outer Approximation Algorithm

The introduction of one or more of the combinatorial constraints presented in Sects. 3.1–3.5 in one of the deterministic formulations equivalent to **PMV** gives a mixed-integer nonlinear programming (MINLP) formulation. Its continuous relaxation is convex and takes the form of a second-order cone programming problem. We illustrate the method with problem **PMVC** that includes diversification, cardinality, buy-in threshold, and concentration constraints. Let $G_{DCL}$ be the mixed-integer set defined as $G_{DCL} = G_D \cap G_C \cap G_L$. Problem **PMVC** reads

$$\textbf{PMVC}: \quad \min \, h$$

$$\text{subject to} \ \ (8); (10); (13); (14)$$

$$w \leq w_{\max} \tag{39}$$

$$\sum_{j \in S_k} w_j \leq L_{\max}, \ l = 1, \ldots, l \tag{40}$$

$$(w, \delta, \zeta) \in \mathcal{X} \cap G_{DCL}. \tag{41}$$

Constraints (39) and (40) are concentration constraints and ensure that a position in an asset (39) or in a sector (40) does not exceed a specified limit $u_{\max}$ or $l_{\max}$.

## 4.1   Algorithm Description

We shall now present an outer approximation algorithm to solve the reformulation of the probabilistic portfolio optimization problem **PMVC** with combinatorial constraints. The proposed algorithm is a derivative of the solution method proposed by [13] and used to construct risk-averse enhanced index funds. The outer approximation algorithm provides a hierarchical organization of the computations in which the set of binary-restricted variables is expanded at each iteration and converges to the exact solution within a finite number of iterations.

At each iteration $t$, we remove the integrality conditions on some of the binary variables $\delta$ and reformulate the cardinality constraint in $G_C$ (21) to obtain an outer approximation of problem **PMVC**. The set $\{1, \ldots, n\}$ of binary variables is partitioned into the subsets $S_1^{(t)}$ and $S_2^{(t)}$, and the integrality restrictions are relaxed on the variables in $S_2^{(t)}$. The partitioning is designed so that the set of variables with integrality restrictions is small enough to make the outer approximation problem easy to solve and large enough to contain, with a high probability, the assets included in the optimal portfolio. The solution of the outer approximation problem gives a lower bound on the optimal solution of the original problem **PMVC** and is used to define the stopping criterion. If the optimal solution of the approximation problem is not feasible for **PMVC**, we tighten the approximation problem for the next iteration $(t + 1)$ by expanding the set of variables on which the binary restrictions are enforced: $S_1^{(t)} \subseteq S_1^{(t+1)}$. Hence, we obtain a series of increasingly tighter outer approximations, which ensures the finite convergence of the algorithm. Note that Coleman et al. [3] have also proposed an outer approximation approach for the solution of deterministic asset allocation problems including a cardinality constraint. While the proposed algorithm solves a finite number of mixed-integer nonlinear programming problems containing a small set of binary variables, Coleman et al. [3] solve a series of continuous outer approximation problems. Our outer approximation formulations are obtained by relaxing the integrality restrictions on some binary variables, whereas Coleman et al. [3] derive continuous outer approximations that proxy the function counting the number of assets in the constructed fund. The next subsections detail the structure of the algorithm. The reader is referred to [1] for a recent and comprehensive review of the available MINLP solvers and algorithms used to solve MINLP problems.

## 4.2   Algorithm Structure

The proposed exact outer approximation algorithm involves an initialization (Sect. 4.2.1) and an iterative (Sect. 4.2.2) phases.

### 4.2.1 Initialization

The successive outer approximation problems are obtained by relaxing the integrality requirements on a subset of the binary variables $\delta$. Let us denote by $G_{CL}^r$ the continuous relaxation of the MIP set $G_{CL} = G_C \cap G_L$. The initial outer approximation problem $\mathbf{OA}^{(0)}$:

$$\mathbf{OA}^{(0)}: \quad \min \quad h$$

$$\text{subject to } (8); (10); (13); (14); (39); (40)$$

$$(w, \zeta) \in \mathcal{X} \cap G_D \tag{42}$$

$$(w, \delta) \in \mathcal{X} \cap G_{CL}^r. \tag{43}$$

is a partial continuous relaxation of $\mathbf{PMVC}$. Indeed, we maintain the integrality restrictions on the vector $\zeta$, which is of small dimension equal to the number of considered industrial sectors. The notation $\delta^{*(t)}$ refers to the value taken by $\delta$ in the optimal solution of the outer approximation problem $\mathbf{OA}^{(t)}$. The same notation style will be used for all decision variables.

If the optimal values $\delta$ in $\mathbf{OA}^{(0)}$ are integer, then the optimal solution $(w^{*(0)}, \delta^{*(0)}, \zeta^{*(0)})$ of $\mathbf{OA}^{(0)}$ is optimal for $\mathbf{PMVC}$ too, and we stop. Otherwise, we start the iterative process and partition the set of securities into:

$$S_1^{(0)} = \{j : w_j^{*(0)} > 0, \ j = 1, \ldots, n\} \tag{44}$$

$$S_2^{(0)} = \{j : w_j^{*(0)} = 0, \ j = 1, \ldots, n\}. \tag{45}$$

The integrality restrictions on the binary variables $\delta_j$, $j \in S_1^{(0)}$ are maintained, while they are removed on those in $S_2^{(0)}$. The idea is to maintain the integrality restrictions on the variables associated with the assets in which the investor would hold positions if the cardinality constraint $e^T \delta \leq u_{\max}$ in $G_C$ was absent. Prior experiments (see also [13]) revealed that a security $j$ with variable $w_j$ taking a positive (resp., null) value in the optimal solution of the relaxed problem is likely (resp., unlikely) to be in the optimal portfolio. Stated differently, the optimal solution of $\mathbf{OA}^{(0)}$ provides key information about the structure of the optimal portfolio.

We shall now use the initial outer approximation problem $\mathbf{OA}^{(0)}$ to derive an optimality cut, also called objective level cut [14], for problem $\mathbf{PMVC}$. We construct a portfolio by only using the assets included in the set $S_1^{(0)}$ whose composition (44) is determined by the optimal solution of $\mathbf{OA}^{(0)}$. This portfolio is obtained by the nonlinear optimization problem $\mathbf{IA}$:

$$\mathbf{IA}: \quad \min \quad h$$

$$\text{subject to } (8); (10); (13); (14); (39); (40); (41)$$

$$w_j = \delta_j = 0, \ j \in S_2^{(0)}. \tag{46}$$

Problem **IA** is easier to solve than **PMVC**, since the continuous relaxation of **IA** is also a second-order cone programming problem, but **IA** has many of its decision variables fixed (46). The asset universe of problem **IA** is restricted to the assets included in $S_1^{(0)}$ and is a subset of the original asset universe used for problem **PMVC**. Proposition 2 follows immediately.

**Proposition 2.** *The nonlinear optimization problem* **IA** *is an inner approximation of problem* **PMVC**. *Let* $u_{IA}^*$ *be the optimal value of* **IA**. *The inequality*

$$h \leq u_{IA}^* \tag{47}$$

*is an objective level cut for* **PMVC**.

Evidently, every feasible solution for **IA** is feasible for **PMVC** and the optimal value $u_{IA}^*$ of **IA** is an upper bound on the optimal value of **PMVC**. Therefore, the cut (47) can eliminate feasible, yet nonoptimal solutions for **PMVC**. As demonstrated in [13], the introduction of such a cut is highly beneficial in speeding up the finding of the optimal solution.

### 4.2.2 Iterative Process

The optimal solution of the outer approximation problem $\mathbf{OA}^{(t-1)}$ is used to update the composition of the sets $S_1^{(t)}$ and $S_2^{(t)}$ at the current iteration $t$:

$$S_1^{(t)} = S_1^{(t-1)} \bigcup \left\{ j : w_j^{*(t-1)} > 0, j \in S_2^{(t-1)} \right\} \tag{48}$$

$$S_2^{(t)} = \{1, 2, \ldots, n\} \setminus S_1^{(t)}. \tag{49}$$

The formulation of the current outer approximation problem $\mathbf{OA}^{(t)}$ is based on the set updating process (48) and (49). Problem $\mathbf{OA}^{(t)}$ is such that (i) a binary variable $\delta_j$ is associated with each asset $j$ in $S_1^{(t)}$ (see (51) and (56)); (ii) there is no binary variable individually assigned to any of the assets in $S_2^{(t)}$. Instead, we associate a binary variable $\delta_S$ with the group of assets in $S_2^{(t)}$ [see (52) and (57)]: $\delta_S$ takes value 1 if any variable $w_j, j \in S_2^{(t)}$ is strictly positive.

$$\mathbf{OA}^{(t)}: \quad \min \quad h \tag{50}$$

$$\text{subject to } (3); (8); (10); (13); (14); (39); (40); (42)$$

$$w_j \leq \delta_j, \ j \in S_1^{(t)} \tag{51}$$

$$\sum_{j \in S_2^{(t)}} w_j \leq \delta_S \tag{52}$$

$$\sum_{j \in S_1^{(t)}} \delta_j + \delta_S \leq u_{max} \tag{53}$$

$$w_{min}\delta_j \leq w_j, \ j \in S_1^{(t)} \tag{54}$$

$$w_{min}\delta_S \leq \sum_{j \in S_2^{(t)}} w_j \tag{55}$$

$$\delta_j \in \{0, 1\}, \ j \in S_1^{(t)} \tag{56}$$

$$\delta_S \in \{0, 1\}. \tag{57}$$

The approximation problem $\mathbf{OA}^{(t)}$ contains $(|S_1^{(t)}| + 1 + l)$ binary variables, while the original problem contains $(n + l)$ of them. In the next $(t + 1)$ iteration, the assets $j$ for which $w_j^{(t)*} > 0$ are moved to $S_1^{(t+1)}$ and a new binary variable is introduced for each of them.

As demonstrated in [13], the outer approximation algorithm outlined above exhibits the following properties.

**Proposition 3.** *Problem* $(\mathbf{OA})^{(t)}$ *is an outer approximation of problem* **PMVC**.

This can be easily seen by observing that the set $\{(w, \delta, \delta_S) : (51)–(57)\}$ is a relaxation of the set $G_{CL}$.

**Proposition 4.** *The optimal solution* $(w^{*(t)}, \delta^{*(t)}, \delta_S^{*(t)}, \zeta^{*(t)})$ *of* $\mathbf{OA}^{(t)}$ *is optimal for* **PMVC** *if*

$$|Q| \leq u_{max}, \quad with \quad Q = \{j : w_j^{*(t)} > 0, j = 1, \ldots, n\}. \tag{58}$$

The above condition (58) is used as the stopping criterion for the algorithmic process.

**Proposition 5.** *The outer approximation algorithm generates a series of increasingly tighter outer approximations.*

It follows immediately from the updating process of the sets $S_1^{(t)}$ and $S_2^{(t)}$ with (48) and (49). At $t$, all the integer variables in $S_1^{(t-1)}$ defined as binary at $(t - 1)$ remain defined as binary, while at least one of the integer variables in $S_2^{(t-1)}$ on which the integrality restriction was relaxed at $(t - 1)$ is binary at $t$.

**Proposition 6.** *The outer approximation algorithm has the finite convergence property. It terminates after at most* $(n\text{-}u_{max}\text{-}1)$ *iterations.*

If $|S_1^{(0)}| \leq u_{max}$, the solution is optimal for **PMVC**. If $|S_1^{(0)}| > u_{max}$, another iteration is needed, in which the integrality restrictions are restored on at least one of the $\delta_j, j \in S_2^{(0)}$ that were so far defined as continuous. Thus, it is clear that the algorithm finds the optimal solution of **PMVC** in at most $(n\text{-}u_{max}\text{-}1)$ iterations. Each iteration consists in the solution of a second-order cone optimization problem with binary variables that optimization solvers such as Cplex or Gurobi can solve with algorithms based on interior point methods and branch-and-bound techniques.

Proposition 5 indicates that each new outer approximation problem has a larger number of binary variables and is likely more challenging to solve. For the algorithm to be efficient, it is crucial that the number of iterations remains small. This underlines the importance of the selection of the variables for which integrality restrictions are enforced. The pseudo-code of the outer approximation algorithm follows.

---

### Pseudo-Code of Outer Approximation Algorithm

---

**Initialization:**
$t := 0$;
Solve the continuous relaxation $\mathbf{OA}^{(0)}$ of **PMVC**. Let
$(w^{*(0)}, \delta^{*(0)}, \zeta^{*(0)}) := argmin(\mathbf{OA}^{(0)})$;
Define $S_1^{(0)}$ and $S_2^{(0)}$ according to (44) and (45);
**if** $|S_1^{(0)}| \leq u_{max}$ **then**
$\quad |\quad (w^{*(0)}, \delta^{*(0)}, \zeta^{*(0)})$ is optimal for **PMVC**
**end**
**else**
$\quad$ Construct and solve inner approximation problem **IA**;
$\quad$ Generate optimality cut (47) and insert it in $\mathbf{OA}^{(0)}$;
$\quad$ **Iterative Process:**
$\quad$ **repeat**
$\quad\quad |\quad t := t + 1$;
$\quad\quad |\quad$ Let $(w^{*(t-1)}, \delta^{*(t-1)}, \delta_S^{*(t-1)}, \zeta^{*(t-1)}) := argmin(\mathbf{OA}^{(t-1)})$;
$\quad\quad |\quad$ Update $S_1^{(t)}$ and $S_2^{(t)}$ as in (48) and (49), respectively ;
$\quad\quad |\quad$ Solve $\mathbf{OA}^{(t)}$;
$\quad$ **until** $|Q| \leq u_{max}$;
**end**

---

## 5 Computational Results

### 5.1 Testbed

For the period spanning from January 1999 to December 2010, we have collected the monthly returns of more than 2,000 stocks from the CRSP US Monthly Stock Database available through the Wharton Research Data Service. Each stock is traded on NYSE and NASDAQ and does not have any missing observation for its

**Table 1** Computational results

| $n$ | ATP CP | ATP OA | NP CP | NP OA | NF CP | NF OA | AIG CP | AIG OA | AOG CP | AOG OA | NB CP | NB OA | NFA CP | NFA OA |
|---|---|---|---|---|---|---|---|---|---|---|---|---|---|---|
| 250 | 208.5 | 187.6 | 8 | 8 | 8 | 8 | 0 | 0 | 0 | 0 | 8 | 8 | 2 | 7 |
| 500 | 1,660.9 | 1,270.1 | 6 | 6 | 7 | 8 | 0.150 % | 0.023 % | 0.008 % | 0 | 7 | 8 | 2 | 6 |
| 750 | 2,387.6 | 1,998.5 | 5 | 6 | 6 | 8 | 0.205 % | 0.068 % | 0.024 % | 0 | 6 | 8 | 0 | 8 |
| 1,000 | 3,222.6 | 2,797.3 | 2 | 4 | 3 | 7 | 0.621 % | 0.157 % | 0.165 % | 0.004 % | 4 | 8 | 1 | 7 |

return data over the considered period. We use the data item called "monthly price alternate" that provides the last available stock price of the month and accounts for splits and dividends. We have built four families of problem instances that differ in the size $n$ (i.e., $n = 250, 500, 750$, and $1,000$) of the market universe. For each family, we have generated eight problem instances. For each of them, the securities included in the asset universe have been selected randomly. We have arbitrarily fixed the number $u_{max}$ of assets that can be in the portfolio to 20.

To evaluate the computational benefits of the proposed algorithm, we have benchmarked the proposed algorithmic OA with the default branch-and-bound algorithm (and default options) of the Cplex 12.4 solver, referred to as CP. Each problem instance has been modeled with the AMPL modeling language and solved with both the OA and CP algorithms on a 64-bit Dell Optiplex 990 Workstation with Quad Core Intel Processor i7-2600 3.40GHz CPU and 16GB of RAM. We shall now use the recorded information to analyze the quality of the solution obtained with and the computational tractability of the proposed hierarchical outer approximation algorithm.

## 5.2 Algorithmic Efficiency

For each family of problem instance, Table 1 displays the average time $ATP$ (in CPU seconds) to prove optimality, the number $NP$ of instances for which optimality is proven, the number $NF$ of instances for which the optimal solution is reached, the average integrality $AIG$ and optimality $AOG$ gaps, the number $NB$ of instances for which the algorithm obtains the best solution, and the number $NFA$ of instances for which the algorithm is the fastest to find the best solution. We define the optimality gap $OG$ (resp., integrality gap $IG$) as the normalized difference between the best solution $z^b$ found in our hour and the known optimal solution $z^*$ (resp., best lower bound $L$):

$$OG = \frac{z^b - z^*}{z^b} \quad \text{and} \quad IG = \frac{z^b - L}{z^b} .$$

Since we solve a minimization problem, and the lower bound is the solution of a relaxation of this problem, we have that $IG \geq OG$. The complete results are given in Table 2 in Appendix. One hour of CPU time is allowed for the solution of each problem. If the optimal solution is not reached within one hour, we set the time equal to 3600 CPU seconds. If the time differential between the two algorithms is less than one second, we consider them equally fast. An entry with 0 in Tables 1 and 2 means that the gap is below $10^{-6}$. In order to obtain some of the results displayed in Table 1, the knowledge of the optimal solution for each instance is needed. We obtained it by letting the algorithms running for more than an hour when required.

It appears that the outer approximation algorithm is, on average, at least 10% faster than CP for each family of problem instances. The OA algorithm is the fastest to reach the best solution for 87.5% of the problem instances. The OA algorithm also outperforms CP in terms on the number of instances for which (i) optimality is proven, and (ii) the optimal solution is reached. It is therefore not surprising that the average optimality and integrality gaps are smaller with OA than with CP. Columns 12 and 13 show that OA finds a better solution than CP for 21.875% of the instances and, in particular, for 50% of the largest instances with 1,000 assets.

Clearly, the proposed outer approximation algorithm permits a more efficient and faster solution of the analyzed instances. It is due to its ability to detect early the structure of the problem and, more precisely, to identify the assets that are likely to be included in the optimal portfolio. Indeed, the outer approximation algorithm requires two or more iterations for only two of the thirty-two problem instances.

## 6  Conclusion

We have studied a probabilistic portfolio optimization problem in which trading restrictions modeled with combinatorial constraints are accounted for. We have provided five deterministic reformulations equivalent to this stochastic programming optimization problem and have discussed their computational efficiency. We have used an exact outer approximation algorithm to solve multiple instances of the reformulated MINLP problem. The algorithm is based on the early recognition of the problem structure and permits a hierarchical organization of the computations. Computational tests show the contribution of the propose algorithm which outperforms the Cplex 12.4 solver in terms of computational times and quality of the obtained solutions. The algorithmic procedure can be easily extended to other asset allocation models with combinatorial constraints.

# Appendix

Table 2 Computational results for thirty-two problem instances

| | Time in CPU seconds | | Integrality gap | | Optimality gap | |
|---|---|---|---|---|---|---|
| n | CP | OA | CP | OA | CP | OA |
| 250-1 | 246.4 | 259.3 | 0 | 0 | 0 | 0 |
| 250-2 | 158.2 | 157.6 | 0 | 0 | 0 | 0 |
| 250-3 | 331.3 | 274.1 | 0 | 0 | 0 | 0 |
| 250-4 | 123 | 118 | 0 | 0 | 0 | 0 |
| 250-5 | 111.2 | 89.6 | 0 | 0 | 0 | 0 |
| 250-6 | 198.6 | 130.9 | 0 | 0 | 0 | 0 |
| 250-7 | 220.7 | 189.3 | 0 | 0 | 0 | 0 |
| 250-8 | 278.5 | 282.3 | 0 | 0 | 0 | 0 |
| 500-1 | 612.7 | 207.6 | 0 | 0 | 0 | 0 |
| 500-2 | 1,128.6 | 441.8 | 0 | 0 | 0 | 0 |
| 500-3 | 2,456.3 | 737.9 | 0 | 0 | 0 | 0 |
| 500-4 | 546.9 | 214.9 | 0 | 0 | 0 | 0 |
| 500-5 | 3,600 | 3,600 | 0.369% | 0.087% | 0 | 0 |
| 500-6 | 897.2 | 904.1 | 0 | 0 | 0 | 0 |
| 500-7 | 3,600 | 3,600 | 0.833% | 0.097% | 0.066% | 0 |
| 500-8 | 445.3 | 454.1 | 0 | 0 | 0 | 0 |
| 750-1 | 3,600 | 2,987.2 | 0.572% | 0 | 0.140% | 0 |
| 750-2 | 1,065.3 | 1,008.3 | 0 | 0 | 0 | 0 |
| 750-3 | 1,048.9 | 642.8 | 0 | 0 | 0 | 0 |
| 750-4 | 1,400.3 | 781.1 | 0 | 0 | 0 | 0 |
| 750-5 | 2,759.8 | 1,902.3 | 0 | 0 | 0 | 0 |
| 750-6 | 3,600 | 3,600 | 0.768% | 0.286% | 0.050% | 0 |
| 750-7 | 3,600 | 3,600 | 0.298% | 0.255% | 0 | 0 |
| 750-8 | 2,026.3 | 1,466.2 | 0 | 0 | 0 | 0 |
| 1,000-1 | 3,600 | 2,719.1 | 0.881% | 0 | 0.368% | 0 |
| 1,000-2 | 3,600 | 3,600 | 0.572% | 0.292% | 0 | 0 |
| 1,000-3 | 3,600 | 3,600 | 0.897% | 0.343% | 0.223% | 0 |
| 1,000-4 | 2,458.3 | 2,473.2 | 0 | 0 | 0 | 0 |
| 1,000-5 | 3,600 | 1,597.1 | 0.832% | 0 | 0.259% | 0 |
| 1,000-6 | 3,600 | 3,600 | 0.891% | 0.361% | 0.396% | 0 |
| 1,000-7 | 3,600 | 3,600 | 0.897% | 0.259% | 0.076% | 0.031% |
| 1,000-8 | 1,722.3 | 1,189.1 | 0 | 0 | 0 | 0 |

# References

1. Bonami, P., Kılınç, M., Linderoth, J.: Algorithms and software for convex mixed integer nonlinear programs. IMA Vol. Math. Appl. 154, 1–39 (2012)
2. Bonami, P., Lejeune, M.A.: An exact solution approach for portfolio optimization problems under stochastic and integer constraints. Oper. Res. 57(3), 650–670 (2009)

3. Coleman, T.F., Li, Y., Henniger, J.: Minimizing tracking error while restricting the number of assets. J. Risk **8**, 33–56 (2006)
4. Cornuejóls, G., Tütüncü, R.: Optimization Methods in Finance. University Press, Cambridge (2007)
5. Dentcheva, D., Ruszczyński, A.: Portfolio optimization with stochastic dominance constraints. J. Bank. Financ. **30**(2), 433–451 (2006)
6. Fabozzi, F.J., Kolm, P.N., Pachamanova, D., Focardi, S.M.: Robust Portfolio Optimization and Management. Wiley, Hoboken, New Jersey, USA (2007)
7. Filomena, T., Lejeune, M.A.: Stochastic portfolio optimization with proportional transaction costs: convex reformulations and computational experiments. Oper. Res Lett. **40**(1), 207–212 (2012)
8. Filomena, T., Lejeune, M.A.: Warm-start heuristic for stochastic portfolio optimization with fixed and proportional transaction costs. J. Optim. Theorey Appl. http://link.springer.com/article/10.1007%2Fs10957-013-0348-y (2013)
9. Guerard, J.B. Jr.: Handbook of Portfolio Construction - Contemporary Applications of Markowitz Techniques. Springer (2010)
10. Jobst, N.J., Horniman, M.D., Lucas, C.A., Mitra, G.: Computational aspects of alternative portfolio selection models in the presence of discrete asset choice constraints. Quant. Financ. **1**, 1–13 (2001)
11. Kataoka, S.: A stochastic programming model. Econometrica **31**(1–2), 181–196 (1963)
12. Lejeune, M.A.: A VaR Black-Litterman Model for the construction of absolute return fund-of-funds. Quant. Financ. **11**, 1489–1501 (2011)
13. Lejeune, M.A., Samatli-Paç, G.: Construction of risk-averse enhanced index funds. INFORMS J. Comput. http://joc.journal.informs.org/content/early/2012/11/26/ijoc.1120.0533.abstract (2013)
14. Li, D., Sun, X.L., Wang, J.: Optimal lot solution to cardinality constrained mean-variance formulation for portfolio selection. Math. Financ. **16**(1), 83–101 (2006)
15. Lobo, M.S., Fazel, M., Boyd, S.: Portfolio optimization with linear and fixed transaction costs. Ann. Oper. Res. **152**, 341–365 (2007)
16. Maringer, D.: Transaction costs and integer constraints. In: Amman, H.M., Rustem B., (eds.) Portfolio Management with Heuristic Optimization - Advances in Computational Management Science. Springer, Dordrecht, The Netherlands **8**, 77–99 (2005)
17. Markowitz, H.M.: Portfolio selection. J. Financ. **7**, 77–91 (1952)
18. Mitra, G., Ellison, F., Scowcroft, A. Quadratic programming for portfolio planning: Insights into algorithmic and computation issues. Part II: processing of portfolio planning models with discrete constraints. J. Asset Manag. **8**, 249–258 (2007)
19. Muthuraman, K., Sunil, K.: Multidimensional portfolio optimization with proportional transaction costs. Math. Financ. **16**, 301–335 (2006)
20. Roy, A.D.: Safety first and the holding assets. Econometrica **20**, 431–449 (1952)
21. Schultz, R., Tiedemann, S.: Risk aversion via excess probabilities in stochastic programs with mixed-integer recourse. SIAM J. Optim. **14**, 115–138 (2006)
22. Welsch, R.E., Zhou, X.: Application of robust statistics to asset allocation models. Revstat **5**, 97–114 (2007)

# An Initialization Strategy for High-Dimensional Surrogate-Based Expensive Black-Box Optimization

Rommel G. Regis

**Abstract** Surrogate-based optimization methods build surrogate models of expensive black-box objective and constraint functions using previously evaluated points and use these models to guide the search for an optimal solution. These methods require considerably more computational overhead and memory than other optimization methods, so their applicability to high-dimensional problems is somewhat limited. Many surrogates, such as radial basis functions (RBFs) with linear polynomial tails, require a maximal set of affinely independent points to fit the initial model. This paper proposes an initialization strategy for surrogate-based methods called *underdetermined simplex gradient descent (USGD)* that uses underdetermined simplex gradients to make progress towards the optimum while building a maximal set of affinely independent points. Numerical experiments on a 72-dimensional groundwater bioremediation problem and on 200-dimensional and 1000-dimensional instances of 16 well-known test problems demonstrate that the proposed USGD initialization strategy yields dramatic improvements in the objective function value compared to standard initialization procedures. Moreover, USGD initialization substantially improves the performance of two optimization algorithms that use RBF surrogates compared to standard initialization methods on the same test problems.

**Keywords** Engineering optimization • High-dimensional black-box optimization • Simplex gradient • Surrogate model • Function approximation • Radial basis function • Expensive function

R.G. Regis (✉)
Department of Mathematics, Saint Joseph's University, Philadelphia, PA 19131, USA
e-mail: rregis@sju.edu

L.F. Zuluaga and T. Terlaky (eds.), *Modeling and Optimization: Theory and Applications*,
Springer Proceedings in Mathematics & Statistics 62, DOI 10.1007/978-1-4614-8987-0__3,

# 1 Introduction

Many engineering optimization problems involve black-box functions whose values are outcomes of computationally expensive simulations. They can be found in engineering design problems in the aerospace and automotive industries and also in parameter estimation problems for mathematical models describing complex physical, chemical, and biological systems. These problems are challenging because only a relatively small number of function evaluations can be used to search for an optimal solution, and they are even more challenging when there are large numbers of decision variables and constraints. Hence, surrogates such as kriging models, radial basis functions (RBFs), and linear and quadratic models are widely used to solve these problems. However, surrogate-based methods tend to require considerably more computational overhead and memory than other optimization methods, so their applicability to high-dimensional problems with hundreds of decision variables is somewhat limited. Moreover, the ability of surrogates to guide the selection of promising iterates tends to diminish as the problem dimension increases. For instance, kriging-based methods have mostly been applied to problems with less than 15 decision variables. However, there are practical optimization problems that involve a large number of decision variables and constraints (e.g., see Jones [34]). This paper proposes an initialization strategy for high-dimensional surrogate-based optimization called *underdetermined simplex gradient descent (USGD)* that uses underdetermined simplex gradients to make progress towards the optimum while building a set of initial function evaluation points. Numerical results show that the proposed USGD strategy generally yields much better objective function values than standard initialization methods on a 72-dimensional groundwater bioremediation management problem and on 200-dimensional and 1000-dimensional instances of 16 well-known test problems. This paper also explores the performance of two surrogate-based optimization algorithms initialized by USGD in comparison to the same algorithms initialized by standard methods on the same problems. The results also show that USGD substantially improves the performance of two optimization algorithms that use RBF surrogates in comparison to standard initialization methods. Hence, the proposed initialization technique facilitates the application of black-box optimization methods to problems with a much larger number of decision variables than those typically considered in the related literature.

The focus of this paper is on the bound-constrained black-box optimization problem:

$$\min \ f(x)$$
$$\text{s.t.} \tag{1}$$
$$x \in \mathbb{R}^d, \ a \leq x \leq b$$

Here, $a, b \in \mathbb{R}^d$, where $d$ is large (possibly in the hundreds or thousands), and $f$ is a deterministic, computationally expensive black-box function. Here, *computationally expensive* means that the evaluation of the objective function completely dominates the cost of the entire optimization process. Moreover, *black box* means that analytical expressions for the objective or constraint functions are not available, and in many cases, the function values are obtained from time-consuming computer simulations. In addition, in many practical applications, the derivatives of the objective function are not explicitly available.

There are also many practical optimization applications with black-box constraints that arise from expensive simulations. For example, Jones [34] presented a black-box automotive problem involving 124 decision variables and 68 black-box inequality constraints and Regis [52, 53] developed RBF methods that work well on this problem. In general, it is difficult to deal with hundreds of decision variables and many constraints in the computationally expensive setting where only relatively few function evaluations can be carried out. In fact, for high-dimensional problems, many algorithms that employ surrogates either run out of memory or take an enormous amount of time for each iteration and sometimes they do not even make substantial progress over the starting solution. For simplicity, this paper focuses only on bound-constrained problems with a large number of decision variables. Future work will develop extensions of the methods proposed in this paper to high-dimensional problems with black-box constraints.

Before proceeding, it is important to clarify that "solving" a high-dimensional and expensive black-box optimization problem does not actually mean finding its global optimum since this is not realistic. Global optimization of inexpensive black-box functions with even a moderate number of dimensions is already computationally challenging. In addition, even convergence to a first-order critical point is hard to achieve for high-dimensional black-box problems when the number of function evaluations is severely limited. In general, algorithms that can be proved to converge to a critical point or to a global minimizer should be preferred. However, another important criterion in practice is how well an algorithm performs when given a severely limited computational budget. Hence, in this paper, algorithms that yield the best objective value after a fixed number of function evaluations and for a wide range of computational budgets and test problems are considered better. Thus, in this context, "solving" a problem simply means providing a good objective function value, in comparison with alternatives, when given a fixed computational budget.

The remaining sections are organized as follows. Section 2 provides a review of literature on black-box optimization methods and surrogate-based methods, including methods for high-dimensional problems. Section 3 presents initialization strategies for high-dimensional surrogate-based optimization, including an algorithm that uses underdetermined simplex gradients. Section 4 presents some numerical experiments. Finally, Sect. 5 presents a summary and some conclusions.

## 2   Review of Literature

A natural approach for expensive black-box optimization problems is to use surrogate models or function approximation models for the expensive functions. Commonly used surrogate modeling techniques include linear and quadratic polynomials, kriging or Gaussian process models [19, 59], RBFs [8, 49], neural networks [11, 32], and support vector machines (SVMs) [39, 66]. Kriging is an interpolation method where the observed function values are assumed to be outcomes of a stochastic process. An advantage of kriging interpolation is that there is a natural way to quantify prediction error at points where the black-box function has not been evaluated. A potential disadvantage of kriging is that it is computationally intensive especially in high dimensions since fitting the model requires numerically solving a maximum likelihood estimation problem. Neural networks are also computationally intensive for high-dimensional problems since fitting the model also requires numerically solving a nonlinear least squares problem. In contrast, the RBF model in Powell [49] can be fit by solving a simple linear system with good theoretical properties.

Surrogate-type approaches have been used in optimization for quite some time. For example, linear and quadratic regression models have been used in response surface methodology [44]. Kriging was used by Jones et al. [33] to develop the efficient global optimization (EGO) method where the next iterate is a global maximizer of an expected improvement function. The convergence of EGO was proved by Vazquez and Bect [69] and explored further by Bull [9]. Variants of EGO have been developed by Huang et al. [29] for stochastic black-box systems and by Aleman et al. [2] for IMRT treatment planning. Kriging was also used by Villemonteix et al. [71] to develop a method that uses minimizer entropy to determine new iterates. On the other hand, RBF interpolation was used by Gutmann [25] to develop a global optimization algorithm where the next iterate is a global minimizer of a measure of bumpiness of the RBF model. This method was also shown to converge to a global minimum by Gutmann [25]. Variants of Gutmann's RBF method have been developed by Björkman and Holmström [6], Regis and Shoemaker [55], Holmström [28], Jakobsson et al. [30], Cassioli and Schoen [10], and Regis and Shoemaker [56].

A promising class of derivative-free expensive black-box optimization methods are those that utilize interpolation models within a trust-region framework. These methods are meant for unconstrained optimization, but they can be easily adapted to handle bound constraints. For example, the DFO method [16, 17], UOBYQA [50], and NEWUOA [51] use quadratic models while BOOSTERS [45, 46] and ORBIT [72, 73] use RBFs. Global convergence of a class of derivative-free trust-region methods to first- and second-order critical points is established by Conn, Scheinberg, and Vicente [15].

A widely used approach for derivative-free black-box optimization is pattern search [65], including extensions such as generating set search (GSS) [35] and mesh adaptive direct search (MADS) [1, 3]. MADS is implemented in the NOMAD software [37]. When the problems are computationally expensive, these methods are sometimes combined with surrogates. For example, Booker et al. [7] and Marsden et al. [40] used kriging with pattern search. Conn and Le Digabel [18] used quadratic models in MADS. Le Thi et al. [38] and Rocha et al. [58] used RBFs while Custódio et al. [20] used minimum Frobenius norm quadratic models in pattern search. Moreover, pattern search can also be made more efficient for expensive functions by using simplex gradients obtained from previous function evaluations [21]. Finally, these methods have parallel implementations such as APPSPACK [24, 36], HOPSPACK [48], and PSD-MADS [4].

Another approach for black-box numerical optimization that is very popular in the engineering optimization community is heuristics (e.g., simulated annealing) and nature-inspired algorithms such as evolutionary algorithms (including genetic algorithms, evolution strategies, evolutionary programming, and scatter search) and swarm intelligence algorithms (e.g., particle swarm and ant colony algorithms). Although genetic algorithms are among the popular metaheuristics, some researchers in evolutionary computation have noted that evolution strategies and evolutionary programming are more suitable for continuous numerical optimization. In particular, CMA-ES (evolution strategy with covariance matrix adaptation) [26, 27] and its variants and extensions have performed well in the annual benchmarking competition in the evolutionary computation community that includes comparisons with other derivative-free methods such as NEWUOA [51]. In contrast to pattern search methods and derivative-free trust-region methods, many of these metaheuristics lack theoretical convergence guarantees and might not perform well on computationally expensive problems when given a very limited computational budget. However, as with pattern search methods, many of these metaheuristics can be combined with surrogates, making them competitive with other derivative-free algorithms. For example, kriging has been used in combination with scatter search [22] and a ranking SVM surrogate has been used for CMA-ES [39]. Jin [31] provides a survey on surrogate-assisted evolutionary computation. In addition, these metaheuristics can also be combined with other methods with convergence guarantees. For example, Vaz and Vicente [67, 68] combined particle swarm with pattern search. Finally, as with pattern search methods, many of these metaheuristics are parallelizable, making it possible to use them on large-scale problems.

When the optimization problems are high dimensional, some surrogate-based approaches, particularly those that rely on kriging models, do not seem suitable because they can become computationally prohibitive and require an enormous amount of memory. In particular, most papers on kriging-based methods, including relatively recent ones, involve less than 15 decision variables (e.g., [12, 22, 47, 70]). Hence, in the area of surrogate-based expensive black-box optimization, problems with more than 15 decision variables are sometimes considered high dimensional and problems with hundreds of decision variables are definitely considered large scale.

One approach for high-dimensional expensive black-box problems is to use dimension reduction techniques such as sequential bifurcation [5] before applying any optimization methods. Another approach is parallelization (e.g., [4, 23]).

Shan and Wang [62] provide a survey of some approaches for high-dimensional, expensive, black-box problems. Because many surrogate models such as kriging can be expensive to maintain, it is not surprising that relatively few surrogate-based or model-based methods have been applied to high-dimensional problems. For example, BOOSTERS [45] and DYCORS [57] have been applied to 200-dimensional problems. Moreover, Shan and Wang [62] developed the RBF-HDMR model and tested it on bound-constrained problems with up to 300 decision variables. Regis [52,53] also developed RBF methods that are suitable for problems with expensive black-box constraints and applied it to a 124-dimensional automotive problem with 68 black-box inequality constraints. Note that what these methods have in common is that they all use RBF surrogates, suggesting that RBF interpolation is promising for high-dimensional expensive black-box problems. Hence, pattern search guided by RBF surrogates and RBF-assisted metaheuristics are also potentially promising for high-dimensional expensive black-box optimization.

## 3   Initialization Strategies for High-Dimensional Surrogate-Based Methods

### 3.1   Standard Initialization Methods

Given a starting point $x_0 \in \mathbb{R}^d$, a standard initialization procedure for surrogate- or model-based methods is to use $d + 1$ initial points consisting of $x_0$ and the points obtained by moving along each of the $d$ positive coordinate directions from $x_0$. That is, use the points $\{x_0, x_0 + \Delta e_1, \ldots, x_0 + \Delta e_d\}$, where $\Delta$ is the step size and the vectors $e_1, \ldots, e_d$ form the natural basis for $\mathbb{R}^d$, i.e., $e_i = [0, \ldots, 0, 1, 0, \ldots, 0]^T$, where the 1 is in the $i$th position. The step size is required to satisfy $\Delta \leq 0.5 \min_{1 \leq i \leq d}(b_i - a_i)$ so that if $x_0 + \Delta e_i$ goes outside the bounds, then it can be replaced by $x_0 - \Delta e_i \in [a, b]$. This procedure is referred to as the *static simplex (SS)* initialization procedure.

   In the high-dimensional and computationally expensive setting, some progress towards the optimum can be made by modifying the static simplex procedure so that the center of the algorithm is always moved to the current best point. Below is the pseudo-code for this method, which is referred to as the *dynamic simplex (DS)* initialization procedure.

**Dynamic Simplex (DS) Initialization Procedure**

**Inputs:**

(1)  Function to minimize: $f : [a, b] \rightarrow \mathbb{R}$, where $[a, b] \subseteq \mathbb{R}^d$
(2)  Starting point: $x_0 \in [a, b]$
(3)  Step size: $\Delta \leq 0.5 \min_{1 \leq i \leq d}(b_i - a_i)$

**Outputs:** A set of $d + 1$ affinely independent points $\mathcal{X} = \{x_0, x_1, \ldots, x_d\} \subseteq \mathbb{R}^d$ and their objective function values $\mathcal{F} = \{f(x_0), f(x_1), \ldots, f(x_d)\}$.

1. **(Evaluate Starting Point)** Calculate $f(x_0)$. Initialize $\mathcal{X} = \{x_0\}$ and $\mathcal{F} = \{f(x_0)\}$. Also, set $x_{\text{best}} = x_0$ and $f_{\text{best}} = f(x_0)$.
2. For $k = 1$ to $d$ do

     (2a) **(Select New Point)** Let $x_k = x_{\text{best}} + \Delta e_k$. If $x_k$ goes outside the bounds, reset $x_k = x_{\text{best}} - \Delta e_k$.

     (2b) **(Evaluate Selected Point)** Calculate $f(x_k)$ and update $x_{\text{best}}$ and $f_{\text{best}}$.

     (2c) **(Update Information)** Reset $\mathcal{X} = \mathcal{X} \cup \{x_k\}$ and $\mathcal{F} = \mathcal{F} \cup \{f(x_k)\}$.

     End for.

3. **(Return Outputs)** Return $\mathcal{X}$ and $\mathcal{F}$.

## 3.2   The Simplex Gradient and Other Preliminaries

Let $\mathcal{X} = \{x_0, x_1, \ldots, x_k\}$ be a set of $k + 1 \leq d + 1$ points in $\mathbb{R}^d$ such that the function values $f(x_0), f(x_1), \ldots, f(x_k)$ are known. If the points in $\mathcal{X}$ are affinely independent, then there exists a linear function (an infinite number if $k < d$) that interpolates the data points $\{(x_0, f(x_0)), (x_1, f(x_1)), \ldots, (x_k, f(x_k))\}$. More precisely, if $p(x) = c_0 + c^T x$, where $c = [c_1, \ldots, c_d]^T$, is a linear polynomial in $d$ variables that interpolates these data points, then

$$\begin{bmatrix} 1 & x_0^T \\ 1 & x_1^T \\ \vdots & \vdots \\ 1 & x_k^T \end{bmatrix} \begin{bmatrix} c_0 \\ c_1 \\ \vdots \\ c_d \end{bmatrix} = \begin{bmatrix} f(x_0) \\ f(x_1) \\ \vdots \\ f(x_k) \end{bmatrix}.$$

The numerical stability of this linear interpolation depends on the condition number of the $(k + 1) \times (d + 1)$ interpolation matrix

$$L(\mathcal{X}) := \begin{bmatrix} 1 & x_0^T \\ 1 & x_1^T \\ \vdots & \vdots \\ 1 & x_k^T \end{bmatrix}$$

If $\mathcal{X}$ is affinely dependent, then $\text{cond}(L(\mathcal{X})) = \infty$. On the other hand, if $\mathcal{X}$ is affinely independent, then $\text{cond}(L(\mathcal{X})) < \infty$ and the smaller the value the better is the geometry of the data points for linear interpolation. A comprehensive treatment of the geometry of sample sets of points for interpolation (determined and underdetermined cases) and regression (overdetermined case) in derivative-free optimization can be found in Conn, Scheinberg, and Vicente [13, 14] and in Scheinberg and Toint [60].

The next section presents an algorithm that iteratively constructs a set $\mathcal{X} = \{x_0, x_1, \ldots, x_d\}$ of $d + 1$ affinely independent points in $\mathbb{R}^d$ while making progress in finding a minimum for $f$ and while keeping $\text{cond}(L(\mathcal{X}))$ relatively small. The points in $\mathcal{X}$ are used to initialize a surrogate-based method, so it is important that the resulting data points have good geometry for interpolation. Moreover, since function evaluations are expensive, it is also important to make progress on the optimization process even during this initialization phase. That is, it is desirable to have a good value for $\min_{0 \leq i \leq d} f(x_i)$ while keeping $\text{cond}(L(\mathcal{X}))$ relatively small. This is accomplished by using the concept of a simplex gradient, which is defined next.

Let $\mathcal{X} = \langle x_0, x_1, \ldots, x_k \rangle$ be an ordered set of $k + 1$ affinely independent points in $\mathbb{R}^d$, where $k \leq d$. Define

$$S(\mathcal{X}) := [x_1 - x_0 \quad \cdots \quad x_k - x_0] \in \mathbb{R}^{d \times k} \quad \text{and} \quad \delta(\mathcal{X}) := \begin{bmatrix} f(x_1) - f(x_0) \\ \vdots \\ f(x_k) - f(x_0) \end{bmatrix} \in \mathbb{R}^k.$$

When $k = d$ (the determined case), $S(\mathcal{X})$ is invertible and the *simplex gradient with respect to* $\mathcal{X}$, denoted by $\nabla_s f(\mathcal{X})$, is given by

$$\nabla_s f(\mathcal{X}) = S(\mathcal{X})^{-T} \delta(\mathcal{X}).$$

When $k < d$ (the underdetermined case), the *simplex gradient with respect to* $\mathcal{X}$ is the minimum 2-norm solution to the system

$$S(\mathcal{X})^T \nabla_s f(\mathcal{X}) = \delta(\mathcal{X}),$$

which is given by $\nabla_s f(\mathcal{X}) = S(\mathcal{X})(S(\mathcal{X})^T S(\mathcal{X}))^{-1} \delta(\mathcal{X})$. In this case, $\nabla_s f(\mathcal{X})$ is a linear combination of $x_1 - x_0, x_2 - x_0, \ldots, x_k - x_0$ since $\nabla_s f(\mathcal{X}) = S(\mathcal{X})v$, where $v = (S(\mathcal{X})^T S(\mathcal{X}))^{-1} \delta(\mathcal{X}) \in \mathbb{R}^k$.

The following proposition shows that it is not necessary that $\mathcal{X}$ be an ordered set when defining the simplex gradient.

**Proposition 1.** *Suppose* $\mathcal{X} = \langle x_0, x_1, \ldots, x_k \rangle$ *is an ordered set of* $k + 1$ *affinely independent points in* $\mathbb{R}^d$, *where* $k \leq d$. *Let* $\alpha$ *be a permutation of the indices* $\{0, 1, \ldots, k\}$ *and let* $\mathcal{X}_\alpha = \langle x_{\alpha(0)}, x_{\alpha(1)}, \ldots, x_{\alpha(k)} \rangle$. *Then* $\nabla_s f(\mathcal{X}_\alpha) = \nabla_s f(\mathcal{X})$.

*Proof.* First consider the case where $\alpha(0) \neq 0$. Then $\alpha(j) = 0$ for some index $1 \leq j \leq k$ and $S(\mathcal{X}_\alpha) = [x_{\alpha(1)} - x_{\alpha(0)} \quad \cdots \quad x_{\alpha(k)} - x_{\alpha(0)}]$ can be transformed to $S(\mathcal{X}) = [x_1 - x_0 \quad \cdots \quad x_k - x_0]$ by applying a series of elementary column operations to $S(\mathcal{X}_\alpha)$. To see this, begin by multiplying the $j$th column of $S(\mathcal{X}_\alpha)$ by $-1$. The result is also given by $S(\mathcal{X}_\alpha)M$, where $M$ is the elementary matrix obtained by replacing the $j$th diagonal entry of $I_d$ by $-1$. Next, for each $i = 1, \ldots, k$, $i \neq j$, perform an elementary column operation that consist of adding the $j$th column of $S(\mathcal{X}_\alpha)M$ to the $i$th column and storing the result in the latter column. The result is $S(\mathcal{X}_\beta)$ for some permutation $\beta$ of the indices $\{0, 1, \ldots, k\}$ that fixes 0. That is,

$$S(\mathcal{X}_\alpha)ME_1E_2 \ldots E_{k-1} = [x_{\beta(1)} - x_0 \quad \cdots \quad x_{\beta(k)} - x_0],$$

where $E_1, E_2, \ldots, E_{k-1}$ are the elementary matrices obtained by adding the $j$th column of $I_d$ to the other columns and storing the results in those columns. Finally, $S(\mathcal{X})$ can be obtained by applying a series of column interchanges to $S(\mathcal{X}_\alpha)ME_1E_2 \ldots E_{k-1}$, i.e.,

$$S(\mathcal{X}_\alpha)ME_1E_2 \ldots E_{k-1}P = S(\mathcal{X}),$$

for some permutation matrix $P$.

Let $F = ME_1E_2 \ldots E_{k-1}P$. Then $S(\mathcal{X}_\alpha)F = S(\mathcal{X})$ and $F$ is nonsingular because it is the product of nonsingular matrices. Observe that

$$F^T\delta(\mathcal{X}_\alpha) = (ME_1E_2 \ldots E_{k-1}P)^T\delta(\mathcal{X}_\alpha) = P^T E_{k-1}^T \ldots E_2^T E_1^T M^T\delta(\mathcal{X}_\alpha) = \delta(\mathcal{X}).$$

Hence,

$$\nabla_s f(\mathcal{X}) = S(\mathcal{X})(S(\mathcal{X})^T S(\mathcal{X}))^{-1}\delta(\mathcal{X}) = S(\mathcal{X}_\alpha)F \left( (S(\mathcal{X}_\alpha)F)^T (S(\mathcal{X}_\alpha)F) \right)^{-1}\delta(\mathcal{X})$$

$$= S(\mathcal{X}_\alpha)F \left( F^T S(\mathcal{X}_\alpha)^T S(\mathcal{X}_\alpha)F \right)^{-1}\delta(\mathcal{X}) = S(\mathcal{X}_\alpha)FF^{-1} \left( S(\mathcal{X}_\alpha)^T S(\mathcal{X}_\alpha) \right)^{-1}(F^T)^{-1}\delta(\mathcal{X})$$

$$= S(\mathcal{X}_\alpha)(S(\mathcal{X}_\alpha)^T S(\mathcal{X}_\alpha))^{-1}\delta(\mathcal{X}_\alpha) = \nabla_s f(\mathcal{X}_\alpha).$$

The proof for the case where $\alpha(0) = 0$ is similar and, in fact, simpler because only permutations are involved.                                                                        □

Next, the following well-known result is useful for understanding the initialization strategies that are given in the succeeding sections.

**Proposition 2.** *Let* $\mathcal{X} = \{x_0, x_1, \ldots, x_k\}$ *be a set of* $k + 1 < d + 1$ *affinely independent points in* $\mathbb{R}^d$. *If* $x_{k+1} \notin \mathrm{aff}(\mathcal{X})$ *(the affine hull of the points in* $\mathcal{X}$*), then* $\mathcal{X} \cup \{x_{k+1}\}$ *is also affinely independent.*

## 3.3 Using Underdetermined Simplex Gradients to Initialize Surrogate-Based Optimization Methods

The main idea of the algorithm below is to build a set of $d + 1$ affinely independent points by iteratively adding new points that are not in the affine hull of the previously chosen points. To increase the chances of making progress in finding the minimum of $f$, a new point is sometimes chosen such that the vector from the current best point (in terms of the value of $f$) to this new point makes an acute angle with the negative of the simplex gradient. However, preliminary numerical experiments show that the condition number of $L(\mathcal{X})$ can quickly deteriorate as more points are added to $\mathcal{X}$ in this particular way. Hence, the algorithm proceeds in two phases. In the first phase, the algorithm performs a series of iterations where it adds a new point such that the vector from the current best point to the new point is perpendicular to the affine hull of the previously chosen points. In this phase, the condition number does not deteriorate much. In the second phase, the algorithm iteratively adds a new point such that the vector from the current best point to this new point makes an acute angle with the negative of the simplex gradient at the current best point. Moreover, to keep the condition number of $L(\mathcal{X})$ to a reasonable value, we find a point $\tilde{x}$ that minimizes $\text{cond}(\mathcal{X} \cup \{\tilde{x}\})$ whenever the condition number exceeds a particular threshold.

## Underdetermined Simplex Gradient Descent (USGD)

### Inputs:

(1) Function to minimize: $f : [a, b] \to \mathbb{R}$, where $[a, b] \subseteq \mathbb{R}^d$
(2) Starting point: $x_0 \in [a, b]$
(3) Step size: $\Delta \leq 0.5 \min_{1 \leq i \leq d}(b_i - a_i)$
(4) Number of perpendicular moves: $0 \leq n_p < d$
(5) Acute angles with the negative simplex gradient: $0 < \theta_k < \pi/2$ for $k = n_p + 1, \ldots, d$
(6) Threshold condition number: $\kappa_{\max}$

**Outputs:** A set of $d + 1$ affinely independent points $\mathcal{X} = \{x_0, x_1, \ldots, x_d\} \subseteq \mathbb{R}^d$ and their objective function values $\mathcal{F} = \{f(x_0), f(x_1), \ldots, f(x_d)\}$.

**Step 1. (Evaluate Starting Point)** Calculate $f(x_0)$. Initialize $\mathcal{X} = \{x_0\}$ and $\mathcal{F} = \{f(x_0)\}$. Also, set $x_{\text{best}} = x_0$ and $f_{\text{best}} = f(x_0)$.
**Phase I: Perpendicular Moves**
**Step 2. (Initialize Set of Coordinates)** Set $\mathcal{C} = \{1, \ldots, d\}$.
**Step 3. (Select Points)** For $k = 1$ to $n_p$ do

(3a) **(Select New Point)** Consider the set of points $\mathcal{T} = \bigcup_{j \in \mathcal{C}} \{x_{\text{best}} \pm \Delta e_j\}$ and let $\tilde{x}$ be a point in $\mathcal{T} \cap [a, b]$ that minimizes $\text{cond}(L(\mathcal{X} \cup \{\tilde{x}\}))$. Let $\tilde{j}$ be the element of $\mathcal{C}$ that gave rise to $\tilde{x}$.
(3b) **(Evaluate Selected Point)** Set $x_k = \tilde{x}$ and calculate $f(x_k)$.
(3c) **(Update Information)** Update $x_{\text{best}}$ and $f_{\text{best}}$. Reset $\mathcal{X} = \mathcal{X} \cup \{x_k\}$ and $\mathcal{F} = \mathcal{F} \cup \{f(x_k)\}$. Also, reset $\mathcal{C} = \mathcal{C} \setminus \{\tilde{j}\}$.

End for.
**Phase II: Acute-Angled Moves Guided by the Negative Simplex Gradient**
**Step 4. (Select Points)** For $k = n_p + 1$ to $d$ do

(4a) **(Identify Best Point)** Let $\ell$ be the index such that $x_{\text{best}} = x_\ell$.
(4b) **(Determine Simplex Gradient)** Calculate $\nabla_s f(\mathcal{X})$.
(4c) **(Determine Vectors Orthogonal to Affine Hull of Previous Points)** Determine an orthonormal basis $\{z_1, \ldots, z_{d-k+1}\}$ for $\text{Null}(S(\mathcal{X})^T)$. (Since $\mathcal{X} = \{x_0, x_1, \ldots, x_{k-1}\}$ consists of $k$ affinely independent points, $\dim(\text{Null}(S(\mathcal{X})^T)) = d - k + 1$).
(4d) For $j = 1$ to $d - k + 1$ do

If $\nabla_s f(\mathcal{X}) \neq 0$, then define $y_{k,j} := w_k z_j - \dfrac{\nabla_s f(\mathcal{X})}{\|\nabla_s f(\mathcal{X})\|}$, where $w_k = \tan(\theta_k)$. Else, define $y_{k,j} := z_j$.

End for.

(4e) Consider the set of trial points $\mathcal{T} = \displaystyle\bigcup_{j=1}^{d-k+1} \left\{ x_{\text{best}} + \Delta \dfrac{y_{k,j}}{\|y_{k,j}\|} \right\}$. Let $\tilde{x}$ be a point in $\mathcal{T} \cap [a, b]$ that minimizes $\text{cond}(L(\mathcal{X} \cup \{\tilde{x}\}))$.

(4f) If the minimum condition number of $L(\mathcal{X} \cup \{\tilde{x}\})$ from (4e) exceeds $\kappa_{\max}$, then replace $\tilde{x}$ by a point in $[a, b]$ that minimizes $\text{cond}(L(\mathcal{X} \cup \{\tilde{x}\}))$.

(4g) **(Evaluate Selected Point)** Set $x_k = \tilde{x}$ and calculate $f(x_k)$.

(4h) **(Update Information)** Update $x_{\text{best}}$ and $f_{\text{best}}$. Reset $\mathcal{X} = \mathcal{X} \cup \{x_k\}$ and $\mathcal{F} = \mathcal{F} \cup \{f(x_k)\}$.

End for.

**Step 5. (Return Outputs)** Return $\mathcal{X}$ and $\mathcal{F}$.

The next proposition is used to verify that the angle between the negative simplex gradient and the vector $\Delta(y_{k,j}/\|y_{k,j}\|)$ (i.e., the vector from the current best point to a trial point) in Step (4e) is the specified angle $\theta_k$.

**Proposition 3.** *Let $v, z \in \mathbb{R}^d$ such that $v \neq 0$ and $z^T v = 0$. Moreover, let $w$ be any positive real number and let $u = wz + v$. Then $u \neq 0$ and the angle between $u$ and $v$ is $\tan^{-1}(w\|z\|/\|v\|)$.*

*Proof.* Let $\theta$ be the angle between $u = wz + v$ and $v$. Since $z^T v = 0$, it follows that

$$\|u\|^2 = u^T u = (wz + v)^T(wz + v) = w^2 z^T z + v^T v = w^2 \|z\|^2 + \|v\|^2. \quad (2)$$

Moreover, since $v \neq 0$, it follows that $\|u\|^2 > 0$, and so, $u \neq 0$. From elementary linear algebra, $0 \leq \theta \leq \pi$ and

$$\cos \theta = \frac{u^T v}{\|u\| \|v\|} = \frac{(wz + v)^T v}{\|u\| \|v\|} = \frac{w(z^T v) + v^T v}{\|u\| \|v\|} = \frac{\|v\|^2}{\|u\| \|v\|} = \frac{\|v\|}{\|u\|}.$$

From the previous equation, $\cos \theta > 0$, and so, $0 \leq \theta < \pi/2$. Now from (2),

$$\tan^2 \theta = \sec^2 \theta - 1 = \frac{\|u\|^2}{\|v\|^2} - 1 = \frac{w^2 \|z\|^2 + \|v\|^2}{\|v\|^2} - 1 = \frac{w^2 \|z\|^2}{\|v\|^2}.$$

Note that $\tan \theta > 0$ since $0 \leq \theta < \pi/2$. Moreover, since $w > 0$, it follows that $\tan \theta = w\|z\|/\|v\|$, and so, $\theta = \tan^{-1}(w\|z\|/\|v\|)$. $\square$

In Step 4(d) of USGD, note that $z_j$ and $-\nabla_s f(\mathcal{X})/\|\nabla_s f(\mathcal{X})\|$ are unit vectors. By Proposition 3, $y_{k,j} \neq 0$ and the angle between $y_{k,j}$ and $-\nabla_s f(\mathcal{X})/\|\nabla_s f(\mathcal{X})\|$ is

$$\tan^{-1}\left( \frac{w_k \|z_j\|}{\| -\nabla_s f(\mathcal{X})/\|\nabla_s f(\mathcal{X})\| \|} \right) = \tan^{-1}(w_k) = \theta_k.$$

Hence, the angle between the negative simplex gradient $-\nabla_s f(\mathcal{X})$ and $y_{k,j}/\|y_{k,j}\|$ (the vector from the current best point to the $j$th trial point in Step 4(e)) is also $\theta_k$.

To show that the above algorithm yields $d + 1$ affinely independent points, we first prove the following simple result.

**Proposition 4.** *Suppose* $\mathcal{X} = \langle x_0, x_1, \ldots, x_k \rangle$ *is an ordered set of* $k + 1$ *affinely independent points in* $\mathbb{R}^d$, *where* $k \leq d$. *Let* $\alpha$ *be a permutation of the indices* $\{0, 1, \ldots, k\}$ *and let* $\mathcal{X}_\alpha = \langle x_{\alpha(0)}, x_{\alpha(1)}, \ldots, x_{\alpha(k)} \rangle$. *Then* $\mathrm{Null}(S(\mathcal{X}_\alpha)^T) = \mathrm{Null}(S(\mathcal{X})^T)$.

*Proof.* By definition,

$$\mathrm{Null}(S(\mathcal{X}_\alpha)^T) = \{z \in \mathbb{R}^d \mid z^T(x_{\alpha(i)} - x_{\alpha(0)}) = 0 \text{ for } i = 1, 2, \ldots, k\}.$$

If $\alpha(0) = 0$, then $\mathrm{Null}(S(\mathcal{X}_\alpha)^T) = \mathrm{Null}(S(\mathcal{X})^T)$. Next, suppose $\alpha(0) \neq 0$. Let $z \in \mathrm{Null}(S(\mathcal{X}_\alpha)^T)$. For any $i = 1, \ldots, k$, note that

$$z^T(x_i - x_0) = z^T((x_i - x_{\alpha(0)}) + (x_{\alpha(0)} - x_0)) = z^T(x_i - x_{\alpha(0)}) - z^T(x_0 - x_{\alpha(0)}) = 0.$$

This shows that $z \in \mathrm{Null}(S(\mathcal{X})^T)$. Hence, $\mathrm{Null}(S(\mathcal{X}_\alpha)^T) \subseteq \mathrm{Null}(S(\mathcal{X})^T)$. A similar argument shows that $\mathrm{Null}(S(\mathcal{X})^T) \subseteq \mathrm{Null}(S(\mathcal{X}_\alpha)^T)$. $\qquad\square$

Next, the following result shows that the acute-angled moves guided by the negative simplex gradient result in affinely independent points.

**Proposition 5.** *Let* $\mathcal{X} = \{x_0, x_1, \ldots, x_k\}$ *be a set of* $k + 1 < d + 1$ *affinely independent points in* $\mathbb{R}^d$ *whose function values* $f(x_0), f(x_1), \ldots, f(x_k)$ *are known and let* $x_\ell \in \mathcal{X}$ *be a point with the smallest function value (i.e.,* $f(x_\ell) \leq f(x_i)$ *for all* $i = 0, 1, \ldots, k$). *Moreover, let* $z \in \mathrm{Null}(S(\mathcal{X})^T)$ *with* $z \neq 0$. *Then for any* $\alpha \neq 0$ *and any constant* $\beta$, $x_\ell + (\alpha z + \beta \nabla_s f(\mathcal{X})) \notin \mathrm{aff}(\mathcal{X})$, *and so, the set* $\mathcal{X} \cup \{x_\ell + (\alpha z + \beta \nabla_s f(\mathcal{X}))\}$ *is also affinely independent.*

*Proof.* Suppose $x_\ell + (\alpha z + \beta \nabla_s f(\mathcal{X})) \in \mathrm{aff}(\mathcal{X})$. Then

$$x_\ell + (\alpha z + \beta \nabla_s f(\mathcal{X})) = a_0 x_0 + a_1 x_1 + \ldots + a_k x_k,$$

where $a_0, a_1, \ldots, a_k$ are constants such that $a_0 + a_1 + \ldots + a_k = 1$. Now

$$\alpha z + \beta \nabla_s f(\mathcal{X}) = a_0 x_0 + a_1 x_1 + \ldots + a_k x_k - (a_0 + a_1 + \ldots + a_k) x_\ell,$$

and so,

$$\alpha z = a_0(x_0 - x_\ell) + a_1(x_1 - x_\ell) + \ldots + a_k(x_k - x_\ell) - \beta \nabla_s f(\mathcal{X}).$$

By Proposition 4,

$$z \in \mathrm{Null}(S(\mathcal{X})^T) = \mathrm{Null}(S(\{x_\ell, x_0, x_1, \ldots, x_{\ell-1}, x_{\ell+1}, \ldots, x_k\})^T)$$

$$= \mathrm{Null}([x_0 - x_\ell, x_1 - x_\ell, \ldots, x_{\ell-1} - x_\ell, x_{\ell+1} - x_\ell, \ldots, x_k - x_\ell]^T).$$

Hence, $z$ is a nonzero vector that is perpendicular to $x_i - x_\ell$ for all $i = 0, 1, \ldots, k$, $i \neq \ell$. Since $\nabla_s f(\mathcal{X})$ is a linear combination of the vectors $x_i - x_\ell$ with $i \neq \ell$, it follows that $z^T \nabla_s f(\mathcal{X}) = 0$ and

$$\alpha \|z\|^2 = \alpha z^T z = z^T (\alpha z) = a_0 z^T (x_0 - x_\ell) + a_1 z^T (x_1 - x_\ell) + \ldots + a_k z^T (x_k - x_\ell)$$
$$-\beta z^T \nabla_s f(\mathcal{X}) = 0.$$

This leads to $\alpha \|z\|^2 = 0$, which is a contradiction since $z \neq 0$ and $\alpha \neq 0$. □

By Proposition 5, each trial point in Step 4(e) does *not* belong to aff($\mathcal{X}$). To see this, note that

$$x_{\text{best}} + \Delta \frac{y_{k,j}}{\|y_{k,j}\|} = x_{\text{best}} + \left( \frac{\Delta w_k}{\|y_{k,j}\|} z_j - \frac{\Delta}{\|y_{k,j}\| \|\nabla_s f(\mathcal{X})\|} \nabla_s f(\mathcal{X}) \right) \notin \text{aff}(\mathcal{X}).$$

Hence, USGD generates a set of $d + 1$ affinely independent points.

The minimization of the condition number in Step 4(f) can be carried out by any standard numerical optimization solver. In this study, this is implemented by creating a Matlab function whose input is a point $x \in [a, b] \subseteq \mathbb{R}^d$ and whose output is cond($L(\mathcal{X} \cup \{\tilde{x}\})$), where $\mathcal{X}$ is the set of previously evaluated points. The Fmincon routine from the Matlab Optimization Toolbox (The MathWorks, Inc. [75]) is then used to find a point $x \in [a, b]$ that minimizes cond($L(\mathcal{X} \cup \{\tilde{x}\})$).

Note that Step 4(e) can take a long time, especially for high-dimensional problems. To speed up the algorithm, one modification is to form the set $\mathcal{T}$ in Step 4(e) using only a random sample of the orthonormal basis from Step 4(c). This modified method is referred to as *USGD-Fast*.

The minimization of the condition number in Step 4(f) can also take a long time. The running time can be reduced by allowing the solver to stop after a maximum number of iterations. Another possibility is to find a quick way to update the condition number of $L(\mathcal{X})$ when a new point is added and this will be explored in future work.

## 4 Numerical Comparison of Initialization Strategies

### 4.1 Test Problems

The three initialization strategies (static simplex (SS), dynamic simplex (DS), and USGD) are compared on the 200-D and 1000-D instances of 16 well-known test functions. Table 1 summarizes the characteristics of these test problems. The first twelve test functions in Table 1 are taken from Moré et al. [43] while the remaining four test functions (the Ackley, Rastrigin, Griewank, and Keane functions) are well known in the engineering optimization community. For the Linear Function–Full

**Table 1** Test problems for the computational experiments

| Test function | Domain | Global min value |
|---|---|---|
| Extended Rosenbrock | $[-2, 2]^d$ | 0 |
| Extended Powell Singular | $[-1, 3]^d$ | 0 |
| Penalty Function I | $[-1, 3]^d$ | Unknown |
| Variably Dimensioned | $[-2, 2]^d$ | 0 |
| Trigonometric | $[-1, 3]^d$ | 0 |
| Brown Almost-Linear | $[-2, 2]^d$ | 0 |
| Discrete Boundary Value | $[-3, 3]^d$ | 0 |
| Discrete Integral Equation | $[-1, 3]^d$ | 0 |
| Broyden Tridiagonal | $[-1, 1]^d$ | 0 |
| Broyden Banded | $[-1, 1]^d$ | 0 |
| Linear Function–Full Rank | $[-2, 1]^d$ | 100 (for $d = 200, m = 300$) |
|  |  | 500 (for $d = 1000, m = 1500$) |
| Linear Function–Rank 1 | $[-1, 3]^d$ | 74.6256 (for $d = 200$) |
|  |  | 374.6251 (for $d = 1000$) |
| Ackley | $[-15, 20]^d$ | $-20 - \exp(1)$ |
| Rastrigin | $[-4, 5]^d$ | $-d$ |
| Griewank | $[-500, 700]^d$ | 0 |
| Keane | $[1, 10]^d$ | Unknown |

Rank and Linear Function–Rank 1 problems from Moré et al. [43], the number of residual functions, denoted by $m$, is set to 300 for the 200-D instances and 1500 for the 1000-D instances. Most of the twelve test problems from Moré et al. [43] each have only one local minimum while the last four problems each have a large number of local minima. Note that 200 and 1000 dimensions are considered very high dimensional in surrogate-based optimization since most papers in this area generally deal with problems with dimensions $d < 15$.

The above test problems are not really computationally expensive to evaluate and the different strategies are compared by pretending that these functions are expensive. This is accomplished by keeping track of the best function values obtained by the different algorithms as the number of function evaluations increases. The relative performance of algorithms on the test problems is expected to be similar to their relative performance on truly expensive functions that have the same general surface as these test problems.

## 4.2 Management of Groundwater Bioremediation

The initialization strategies are also compared on a groundwater bioremediation problem [41, 74]. Groundwater bioremediation involves using injection wells that supply electron acceptors (e.g., oxygen) or electron donors (e.g., hydrogen) into the groundwater to promote the growth of the soil bacteria that can transform contaminants into harmless substances. Monitoring wells are also used to measure the concentration of the contaminant at specific locations and ensure that it is below some threshold level at specified time periods.

The setup involves a hypothetical contaminated aquifer whose characteristics are symmetric about a horizontal axis. The aquifer is discretized using a 2-D finite element mesh with 18 nodes in the horizontal direction and 9 nodes in the vertical direction. There are 6 injection wells and 84 monitoring wells that are also symmetrically arranged. Oxygenated water is pumped into the injection wells. The optimization formulation involves a 2-D finite element simulation model that describes groundwater flow and changes in the concentrations of the contaminant, oxygen, and biomass. The entire planning horizon is evenly divided into 24 management periods and the problem is to determine the pumping rates for each injection well at the beginning of each management period in order to minimize the total pumping cost subject to some constraints on the contaminant concentration at the monitoring wells. The problem can be reformulated as a box-constrained global optimization problem by incorporating the constraints into the objective using a penalty term [74]. Because the wells are symmetrically arranged, pumping decisions are only needed for the 3 injection wells on one side of the axis of symmetry resulting in 72 decision variables for the optimization problem. The maximum pumping rate is rescaled to 1, so the search space is $[0, 1]^{72}$. This groundwater bioremediation problem is referred to as GWB72 and it is an extension of the 12-dimensional problem in Regis and Shoemaker [54]. This groundwater bioremediation model uses a relatively coarse grid, so its simulation time is only about 0.1 s on an Intel(R) Core(TM) i7 CPU 860 2.8 GHz desktop machine. However, it is representative of more complex groundwater bioremediation problems whose simulation times can sometimes take several hours [63].

## 4.3   Experimental Setup

Two sets of numerical experiments are performed to assess the effectiveness of the USGD initialization strategy. In the first set of experiments, the USGD strategy is compared with the static simplex (SS) and dynamic simplex (DS) strategies in terms of the mean of the best objective function value. In the second set of experiments, each surrogate-based algorithm is run with each of the three different initialization strategies (SS, DS, and USGD).

For the first set of experiments, each initialization strategy is run for 30 trials on the 72-D groundwater bioremediation problem GWB72 and on each of the 200-D test problems. However, since these strategies can become computationally expensive on the 1000-D test problems, they are only run for 10 trials on these problems. Each trial corresponds to a fixed starting point that is the same for all methods and all strategies use the same step size of $\Delta = 0.2 \min_{1 \leq i \leq d} (b_i - a_i)$. For GWB72 and the 200-D problems, USGD is run with $n_p = \lfloor d/2 \rfloor$, $\theta_k = 75^o$ for all $k$, and $\kappa_{max} = 10^5$. However, for the 1000-D problems, the faster implementation USGD-Fast is run with $n_p = \lfloor 3d/4 \rfloor$, $\theta_k = 80^o$ for all $k$, and $\kappa_{max} = 10^6$. The reason for delaying the acute-angled moves guided by the negative simplex gradient and for using steeper angles for the 1000-D problems is to prevent the condition number of the linear interpolation matrix from becoming very large. The threshold condition numbers $\kappa_{\max} = 10^5$ or $10^6$ are set based on numerical experiments with the condition numbers of the linear interpolation matrices of randomly generated points. More precisely, when $d = 200$, the mean and median condition numbers (out of 10,000 trials) of linear interpolation matrices of uniform random points on $[0, 1]^d$ are $6.28 \times 10^4$ and $9.99 \times 10^3$, respectively. When $d = 1000$, the mean and median condition numbers (out of 10,000 trials) of linear interpolation matrices of uniform random points on $[0, 1]^d$ are $5.68 \times 10^5$ and $1.14 \times 10^5$, respectively. All methods are run in Matlab 7.11 using an Intel(R) Core(TM) i7 CPU 860 2.8 GHz desktop machine.

For the second set of numerical experiments, two surrogate-based methods for high-dimensional, bound-constrained, expensive black-box optimization are run with each of the three initialization strategies (SS, DS, and USGD or USGD-Fast) on GWB72 and on the 200-D and 1000-D test problems. Again, USGD is used for GWB72 and the 200-D problems while USGD-Fast is used for the 1000-D problems. The surrogate-based methods used are DYCORS-DDSRBF [57] and a Matlab implementation of a pattern search algorithm that uses RBF surrogates [38], which is denoted by PS-RBF. The particular RBF model used for both DYCORS-DDSRBF and PS-RBF is a cubic RBF augmented by a linear polynomial tail, which has been used by Björkman and Holmström [28], Wild, Regis, and Shoemaker [72], Regis [52], Le Thi et al. [38], and Regis and Shoemaker [57]. Thus, six methods are compared: DYCORS-DDSRBF (SS), DYCORS-DDSRBF (DS), DYCORS-DDSRBF (USGD or USGD-Fast), PS-RBF (SS), PS-RBF (DS), and PS-RBF (USGD or USGD-Fast). Each combination of optimization method and initialization strategy is run for 30 trials on GWB72 and on the 200-D problems and for only 10 trials on the 1000-D problems. Each trial begins with the initial points generated by the assigned initialization strategy and this is the same for DYCORS-DDSRBF and PS-RBF with this initialization strategy. All computational runs are also carried in Matlab 7.11 using the same machine that was used to test the initialization strategies.

DYCORS-DDSRBF is an RBF-assisted modification of the *DDS (dynamically dimensioned search)* heuristic by Tolson and Shoemaker [64]. It follows the *DYCORS (DYnamic COordinate search using Response Surface models)* framework for bound-constrained, high-dimensional, expensive black-box optimization by Regis and Shoemaker [57]. In the DYCORS framework, a response surface (or surrogate) model is used to select the iterate from a set of random trial solutions generated by perturbing only a subset of the coordinates of the current best solution. Moreover, the probability of perturbing a coordinate of the current best solution decreases as the algorithm progresses. PS-RBF is a pattern search algorithm guided by RBF surrogates [38] that is implemented in the PSwarm solver [67, 68]. In PS-RBF, an RBF model is built and minimized during the search step of the underlying pattern search method. The RBF model is minimized within an $\infty$-norm trust region whose radius is proportional to the step size and the resulting box-constrained problem is solved using a d.c. programming method. In addition, the RBF model is used to sort the directions in the poll step of the pattern search method.

Combining the DYCORS-DDSRBF solver with the initialization strategies is straightforward. The set of affinely independent points and objective function values obtained by each initialization strategy are used as inputs to the solver. The DYCORS-DDSRBF solver then begins by fitting the initial RBF model that is used to select the next iterate. Ideally, merging an initialization strategy with the PS-RBF solver should be done in the same manner. However, the PS-RBF code from the PSwarm solver does not accept a set of points as input. It only accepts a single starting point, so the best point (in terms of objective function value) obtained by an initialization strategy is used as the starting point for PS-RBF. Hence, one limitation of this study is that the PS-RBF solver is not properly merged with the initialization strategies, and any improvement of PS-RBF (USGD or USGD-Fast) over PS-RBF (DS) or PS-RBF (SS) is most likely due only to the fact that the starting point of the former has better objective function value than the starting points of the latter algorithms. The performance of PS-RBF can be further improved if the entire set of points obtained by an initialization strategy is used by the solver.

Before proceeding with the comparisons, it is important to clarify that the purpose of the comparisons below is *not* to demonstrate that one algorithm is superior to the other methods. In particular, it is not the intent of the paper to show that DYCORS-DDSRBF (USGD) is better than PS-RBF (USGD) or even to prove that DYCORS-DDSRBF (USGD) is always better than DYCORS-DDSRBF (SS). First of all, the comments in the previous paragraph suggest that any comparison between DYCORS-DDSRBF and PS-RBF combined with any initialization strategy would not be fair. Second, the test problems are relatively limited, so one cannot really generalize the results to a larger class of problems. After all, it is widely believed that there is no universally best optimization method. Finally, it is difficult to guarantee that each method is run with the best parameter settings for the given problems. Although default and reasonable parameter settings are used for the different methods, it might still be possible to improve their performance by more carefully tuning their parameters (some of which are not visible to the user). The goal is simply to make a case why some of these methods (e.g., DYCORS-DDSRBF (USGD)) should be seriously considered for solving high-dimensional expensive black-box optimization problems.

## 4.4   Results of Underdetermined Simplex Gradient Descent

Table 2 shows the results of applying the USGD initialization procedure and the two other alternatives (SS and DS) on the 200-D test problems. Table 3 shows the results of applying USGD-Fast and the other methods on the 1000-D problems. Note that the simple modification provided by DS already yields large improvements on the best objective function value over SS. However, USGD and USGD-Fast yield even better improvements over DS on 15 of the 16 problems (all except the Keane problem). Moreover, on the Keane problem, DS is only slightly better than USGD or USGD-Fast. These results suggest that moving in the direction that makes an

**Table 2** Mean and standard error of the best objective function values in 30 trials for three initialization strategies for surrogate-based optimization methods on GWB72 and on the 200-D problems

| Test function | USGD | Dynamic simplex (DS) | Static simplex (SS) |
|---|---|---|---|
| GWB72 | 548.78 (18.27) | 1919.57 (37.88) | 2368.24 (52.41) |
| Extended Rosenbrock | 3347.54 (131.95) | 5959.02 (162.63) | 44334.37 (1153.21) |
| Extended Powell Singular | 7342.31 (207.92) | 20917.62 (627.58) | 80834.69 (3313.60) |
| Penalty Function I | 6534.78 (308.14) | 96551.29 (2571.04) | 217584.45 (5761.92) |
| Variably Dimensioned | $1.37 \times 10^{15}$ $(2.08 \times 10^{14})$ | $3.49 \times 10^{15}$ $(4.32 \times 10^{14})$ | $1.64 \times 10^{17}$ $(1.04 \times 10^{16})$ |
| Trigonometric | $4.92 \times 10^5$ $(1.91 \times 10^4)$ | $7.46 \times 10^6$ $(1.72 \times 10^5)$ | $1.18 \times 10^7$ $(2.71 \times 10^5)$ |
| Brown Almost-Linear | $1.21 \times 10^5$ $(1.84 \times 10^4)$ | $1.10 \times 10^6$ $(7.21 \times 10^4)$ | $8.00 \times 10^6$ $(2.24 \times 10^5)$ |
| Discrete Boundary Value | 422.43 (12.02) | 964.61 (21.58) | 3522.71 (65.03) |
| Discrete Integral Equation | 94.75 (2.72) | 752.14 (14.49) | 1176.89 (23.23) |
| Broyden Tridiagonal | 231.48 (4.16) | 327.81 (6.51) | 993.18 (16.51) |
| Broyden Banded | 399.72 (7.59) | 756.48 (10.95) | 3225.48 (53.12) |
| Linear Function–Full Rank | 182.50 (1.83) | 212.83 (1.61) | 299.58 (2.54) |
| Linear Function–Rank 1 | $2.65 \times 10^{12}$ $(1.04 \times 10^{12})$ | $2.59 \times 10^{15}$ $(8.62 \times 10^{13})$ | $3.64 \times 10^{15}$ $(1.21 \times 10^{14})$ |
| Ackley | −9.09 (0.15) | −6.21 (0.05) | −3.50 (0.03) |
| Rastrigin | 156.30 (8.51) | 548.82 (7.82) | 1398.10 (15.73) |
| Griewank | 604.68 (15.62) | 2841.80 (39.97) | 6492.08 (74.85) |
| Keane | −0.1439 (0.0016) | −0.1537 (0.0009) | −0.0895 (0.0014) |

acute angle with the negative simplex gradient during the initialization phase for a surrogate-based optimization method results in significant improvement on the objective function value on high-dimensional problems.

Tables 4 and 5 show the means and standard errors of the condition numbers of the resulting linear interpolation matrices for the three initialization strategies on each of the test problems. The means of the condition numbers for USGD are worse than those for DS on 10 of the 17 problems in Table 4 and on all 16 problems in Table 5. However, they are better than those for SS on 16 of the problems in Table 4 (all except the 200-D Extended Powell Singular) and on 12 of the problems in Table 5. Moreover, the means of the condition numbers for USGD are 481.84 for GWB72 and $< 4 \times 10^4$ on 14 of the 200-D problems (all except the 200-D Griewank and Keane problems), and these are well within the threshold condition number $\kappa_{max} = 10^5$ for GWB72 and the 200-D problems. In addition, the means of the condition numbers for USGD-Fast are $< 5 \times 10^5$ on 14 of the 1000-D problems (all except the 1000-D Griewank and Keane problems), and these are again within the threshold condition number $\kappa_{max} = 10^6$ for the 1000-D problems.

**Table 3** Mean and standard error of the best objective function values in 10 trials for three initialization strategies for surrogate-based optimization methods on the 1000-D problems

| Test function | USGD-Fast | Dynamic simplex (DS) | Static simplex (SS) |
|---|---|---|---|
| Extended Rosenbrock | 26,156.39 (741.33) | 30,285.17 (401.52) | 229,789.24 (3678.87) |
| Extended Powell Singular | 66,230.64 (1146.68) | 107,335.62 (2210.79) | 449,075.45 (10161.50) |
| Penalty Function I | 822,745.49 (18160.28) | 2,394,676.33 (56040.92) | 5,396,250.76 (115621.31) |
| Variably Dimensioned | $3.65 \times 10^{20}$ ($4.11 \times 10^{19}$) | $1.22 \times 10^{21}$ ($1.12 \times 10^{20}$) | $6.54 \times 10^{22}$ ($2.84 \times 10^{21}$) |
| Trigonometric | $2.91 \times 10^{8}$ ($5.55 \times 10^{6}$) | $9.27 \times 10^{8}$ ($1.75 \times 10^{7}$) | $1.47 \times 10^{9}$ ($2.53 \times 10^{7}$) |
| Brown Almost-Linear | $2.73 \times 10^{7}$ ($2.47 \times 10^{6}$) | $1.37 \times 10^{8}$ ($7.65 \times 10^{6}$) | $1.02 \times 10^{9}$ ($2.51 \times 10^{7}$) |
| Discrete Boundary Value | 2534.10 (283.85) | 4993.86 (77.45) | 18506.10 (340.80) |
| Discrete Integral Equation | 1472.63 (28.91) | 3718.43 (60.45) | 5864.37 (103.47) |
| Broyden Tridiagonal | 1277.67 (50.33) | 1671.79 (28.70) | 5271.13 (93.70) |
| Broyden Banded | 3289.52 (44.03) | 3801.36 (31.98) | 16536.81 (154.76) |
| Linear Function–Full Rank | 764.87 (6.93) | 1062.23 (6.85) | 1497.04 (9.56) |
| Linear Function–Rank 1 | $1.28 \times 10^{19}$ ($9.35 \times 10^{17}$) | $1.95 \times 10^{20}$ ($5.04 \times 10^{18}$) | $2.76 \times 10^{20}$ ($6.76 \times 10^{18}$) |
| Ackley | $-8.77$ (0.24) | $-6.19$ (0.04) | $-3.47$ (0.02) |
| Rastrigin | 1890.34 (22.24) | 2722.35 (35.24) | 7013.49 (49.99) |
| Griewank | 6862.44 (98.96) | 14180.55 (169.62) | 32518.19 (239.73) |
| Keane | $-0.1449$ (0.0008) | $-0.1552$ (0.0007) | $-0.0894$ (0.0009) |

## 4.5 Performance and Data Profiles

In the second set of experiments, DYCORS-DDSRBF and PS-RBF initialized by USGD are compared with these same algorithms initialized by SS and DS using performance and data profiles [42]. Let $\mathcal{P}$ be the set of problems where a given problem $p$ corresponds to a particular test problem and a particular starting point (prior to the initialization strategy). The performance and data profiles are created separately for the 200-D and 1000-D test problems. Since there are sixteen 200-D (and also sixteen 1000-D) problems and 30 starting points (corresponding to the 30 trials), there are $16 \times 30 = 480$ problems for the profiles. Moreover, let $\mathcal{S}$ be the set of solvers. Here, there are 6 solvers (DYCORS-DDSRBF (USGD), DYCORS-DDSRBF (DS), DYCORS-DDSRBF (SS), PS-RBF (USGD), PS-RBF (DS), and PS-RBF (SS)). For any pair $(p, s)$ of a problem $p$ and a solver $s$, the *performance ratio* is defined by

$$r_{p,s} := \frac{t_{p,s}}{\min\{t_{p,s} : s \in \mathcal{S}\}},$$

**Table 4** Mean and standard error of the condition numbers of the resulting interpolation matrix in 30 trials for three initialization strategies for surrogate-based optimization methods on GWB72 and on the 200-D problems

| Test function | USGD | Dynamic simplex (DS) | Static simplex (SS) |
|---|---|---|---|
| GWB72 | 481.84 (13.78) | 2012.91 (59.30) | 2816.69 (118.67) |
| Extended Rosenbrock | 17,418.74 (1151.65) | 3766.98 (54.66) | 38,613.17 (1009.10) |
| Extended Powell Singular | 18,529.61 (952.13) | 10,900.17 (157.00) | 13,224.10 (865.08) |
| Penalty Function I | 4173.08 (295.32) | 10,101.70 (179.56) | 13,224.10 (865.08) |
| Variably Dimensioned | 14,442.34 (989.06) | 6481.60 (90.49) | 38,613.17 (1009.10) |
| Trigonometric | 4914.77 (208.29) | 10,110.82 (178.31) | 13,224.10 (865.08) |
| Brown Almost-Linear | 7774.41 (460.17) | 6481.60 (90.49) | 38,613.17 (1009.10) |
| Discrete Boundary Value | 16,309.45 (1329.72) | 6365.37 (120.69) | 57,676.07 (1516.08) |
| Discrete Integral Equation | 4879.38 (212.35) | 10,026.23 (266.24) | 13,224.10 (865.08) |
| Broyden Tridiagonal | 11,248.22 (369.68) | 2568.09 (45.59) | 19,743.66 (500.35) |
| Broyden Banded | 5693.20 (186.77) | 1476.10 (19.41) | 19,743.66 (500.35) |
| Linear Function–Full Rank | 19,626.08 (1265.89) | 4217.09 (89.41) | 53,051.24 (986.62) |
| Linear Function–Rank 1 | 7619.04 (442.15) | 9998.30 (268.55) | 13,224.10 (865.08) |
| Ackley | 31,655.90 (3313.84) | 31,634.32 (500.27) | 259,032.55 (9336.27) |
| Rastrigin | 27,343.19 (2146.34) | 7514.11 (99.70) | 70,841.79 (2351.12) |
| Griewank | 576,644.29 (32668.25) | 1,156,648.70 (17580.00) | 8,477,462.48 (325536.92) |
| Keane | 138,487.01 (12079.53) | 82,492.99 (1160.75) | 245,775.76 (10121.11) |

where $t_{p,s}$ is the number of function evaluations required to satisfy the convergence test that is defined below. Note that $r_{p,s} \geq 1$ for any $p \in \mathcal{P}$ and $s \in \mathcal{S}$. Moreover, for a given problem $p$, the best solver $s$ for this problem attains $r_{p,s} = 1$. Furthermore, by convention, set $r_{p,s} = \infty$ whenever solver $s$ fails to yield a solution that satisfies the convergence test.

Now, for any solver $s \in \mathcal{S}$ and for any $\alpha \geq 1$, the *performance profile of $s$ with respect to $\alpha$* is the fraction of problems where the performance ratio is at most $\alpha$, i.e.,

$$\rho_s(\alpha) = \frac{1}{|\mathcal{P}|} \left| \{ p \in \mathcal{P} : r_{p,s} \leq \alpha \} \right|.$$

For any solver $s \in \mathcal{S}$, the *performance profile curve of $s$* is the graph of the performance profiles of $s$ for a range of values of $\alpha$.

Derivative-free algorithms for expensive black-box optimization are typically compared given a fixed and relatively limited number of function evaluations. In particular, the convergence test by Moré and Wild [42] uses a tolerance $\tau > 0$ and the minimum function value $f_L$ obtained by *any* of the solvers on a particular problem within a given number $\mu_f$ of function evaluations and it checks if a point $x$ obtained by a solver satisfies

$$f(x_0) - f(x) \geq (1 - \tau)(f(x_0) - f_L),$$

**Table 5** Mean and standard error of the condition numbers of the resulting interpolation matrix in 10 trials for three initialization strategies for surrogate-based optimization methods on the 1000-D problems

| Test function | USGD-Fast | Dynamic simplex (DS) | Static simplex (SS) |
|---|---|---|---|
| Extended Rosenbrock | 132,474.48 (12,672.99) | 41,834.45 (372.65) | 942,541.41 (13,273.37) |
| Extended Powell Singular | 297,399.59 (38,993.88) | 120,619.52 (1481.65) | 142,210.10 (11,636.73) |
| Penalty Function I | 244,556.11 (46,266.62) | 113,910.31 (831.45) | 142,210.10 (11,636.73) |
| Variably Dimensioned | 108,801.26 (9040.40) | 72,204.04 (753.03) | 942,541.41 (13,273.37) |
| Trigonometric | 141,579.52 (25,123.47) | 113,903.08 (832.55) | 142,210.10 (11636.73) |
| Brown Almost-Linear | 88,640.16 (6399.78) | 72,204.04 (753.03) | 942,541.41 (13273.37) |
| Discrete Boundary Value | 142,872.55 (16285.88) | 68,829.80 (1205.76) | 1,412,591.21 (19916.05) |
| Discrete Integral Equation | 170,963.99 (19170.73) | 112,284.62 (721.91) | 142,210.10 (11636.73) |
| Broyden Tridiagonal | 69,791.48 (4071.08) | 26,786.60 (348.40) | 473,466.72 (6625.97) |
| Broyden Banded | 91,703.76 (4934.80) | 15,279.28 (165.98) | 473,466.72 (6625.97) |
| Linear Function–Full Rank | 115,467.93 (11176.48) | 46,730.66 (900.64) | 1,320,508.47 (19257.26) |
| Linear Function–Rank 1 | 162,938.86 (20,618.71) | 112,057.14 (926.10) | 142,210.10 (11636.73) |
| Ackley | 453,153.51 (64,588.20) | 355,482.96 (3976.48) | 6,241,984.99 (102133.16) |
| Rastrigin | 203,772.79 (19900.39) | 84,665.57 (915.34) | 1,713,659.81 (26601.73) |
| Griewank | $1.62 \times 10^7$ $(5.99 \times 10^6)$ | $1.30 \times 10^7$ $(1.32 \times 10^5)$ | $2.03 \times 10^8$ $(3.49 \times 10^6)$ |
| Keane | $1.23 \times 10^6$ $(9.84 \times 10^4)$ | $9.17 \times 10^5$ $(9.93 \times 10^3)$ | $6.23 \times 10^6$ $(9.33 \times 10^4)$ |

where $x_0$ is a starting point corresponding to the problem under consideration. In the above expression, $x$ is required to achieve a reduction that is $1 - \tau$ times the best possible reduction $f(x_0) - f_L$.

Next, given a solver $s \in \mathcal{S}$ and $\alpha > 0$, the *data profile of a solver s with respect to $\alpha$* [42] is given by

$$d_s(\alpha) = \frac{1}{|\mathcal{P}|} \left| \left\{ p \in \mathcal{P} : \frac{t_{p,s}}{n_p + 1} \leq \alpha \right\} \right|,$$

where $t_{p,s}$ is the number of function evaluations required by solver $s$ to satisfy the convergence test on problem $p$ and $n_p$ is the number of variables in problem $p$.

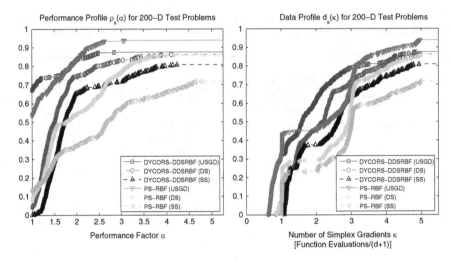

**Fig. 1** Performance and data profiles for the alternative optimization methods on the 200-D test problems

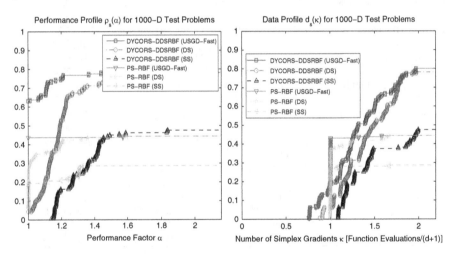

**Fig. 2** Performance and data profiles for the alternative optimization methods on the 1000-D test problems

For any solver $s \in \mathcal{S}$, the *data profile curve of s* is the graph of the data profiles of $s$ for a range of values of $\alpha$. For a given solver $s$ and any $\alpha > 0$, $d_s(\alpha)$ is the fraction of problems "solved" (i.e., problems where the solver generated a point satisfying the convergence test) by $s$ within $\alpha(n_p + 1)$ function evaluations (equivalent to $\alpha$ simplex gradient estimates [42]).

Figures 1 and 2 show the performance profile and data profiles [42] for the 200-D and 1000-D problems. For the 200-D problems, all algorithms are run up to 1000 function evaluations, while for the 1000-D problems, the algorithms are run up to 2000 function evaluations. Recall, however, that the first 201 function evaluations for the 200-D problems and the first 1001 function evaluations on the 1000-D problems are spent on the initialization procedure (USGD, DS, or SS). The parameter $\tau$ for the performance and data profiles described above is set to $\tau = 0.05$.

The performance and data profiles show that DYCORS-DDSRBF (USGD) is generally much better than both DYCORS-DDSRBF (SS) and DYCORS-DDSRBF (DS) on the 200-D and 1000-D test problems. Moreover, PS-RBF (USGD) is generally much better than both PS-RBF (SS) and PS-RBF (DS) on the 200-D problems and it is also generally better than PS-RBF (SS) on the 1000-D problems. The profiles for the 1000-D problems do not show any clear improvement of PS-RBF (USGD) over PS-RBF (DS). However, USGD does not seem to hurt the performance of PS-RBF on the 1000-D problems. Hence, these results suggest that the USGD initialization strategy is generally helpful in improving the performance of DYCORS-DDSRBF and, to a limited extent, of PS-RBF. These results also suggest that USGD is potentially helpful for other surrogate-based methods, especially those that require a maximal set of affinely independent points for initialization.

Recall, however, that the points generated by the various initialization strategies (USGD, DS, and SS) are not fully integrated into PS-RBF as explained in Sect. 4.3. PS-RBF only uses the best point from the set of points generated by these initialization strategies and ignores the rest of the points. Hence, the improvements of PS-RBF (USGD) over either PS-RBF (SS) or PS-RBF (DS) are mostly due to the fact that the starting point of PS-RBF after USGD typically has a much better objective function value than the starting point of PS-RBF after either SS or DS. In contrast, DYCORS-DDSRBF uses all the points generated by these initialization strategies. Because of this, it would not be fair to compare DYCORS-DDSRBF and PS-RBF in this paper.

## 4.6   Average Progress Curves

The different combinations of surrogate-based method and initialization strategy are also compared using average progress curves. An *average progress curve* is a graph of the mean of the best objective function value obtained by an algorithm as the number of function evaluations increases. Figures 3–10 show the average progress curves when the various optimization algorithms are applied to the 72-D groundwater bioremediation problem GWB72 and to the 200-D and 1000-D test problems. Figures 8–10 are in the appendix. The error bars in these figures represent 95% t-confidence intervals for the mean. That is, each side of the error bar has length equal to 2.045 (for 30 trials) or 2.262 (for 10 trials) times the standard deviation of the best function value divided by the square root of the number of trials. Here, 2.045 and 2.262 are the critical values corresponding to a 95% confidence level for a t distribution with 29 and 9 degrees of freedom, respectively.

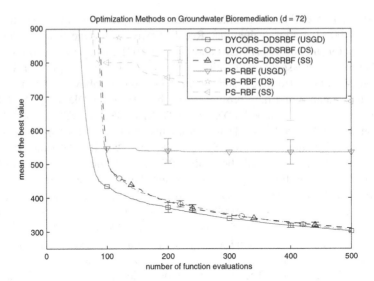

**Fig. 3** Mean of the best objective function value (over 30 trials) vs number of function evaluations for the alternative optimization methods on the 72-D groundwater bioremediation problem. Error bars represent 95% t-confidence intervals for the mean

Figure 3 shows that DYCORS-DDSRBF (USGD) is better than both DYCORS-DDSRBF (SS) and DYCORS-DDSRBF (DS) on GWB72. Moreover, PS-RBF (USGD) is also better than both PS-RBF (SS) and PS-RBF (DS) on GWB72. However, multiple trials of PS-RBF (USGD) appear to be stuck at a local minimum relatively early in the search.

Figures 4–7 show that DYCORS-DDSRBF (USGD) is better than both DYCORS-DDSRBF (SS) and DYCORS-DDSRBF (DS) on 13 of the 200-D test problems (all except Variably Dimensioned, Broyden Tridiagonal, and Brown Almost-Linear). Moreover, the performance of DYCORS-DDSRBF (USGD) is comparable to that of the latter algorithms on the 200-D Variably Dimensioned and Broyden Tridiagonal problems. Note that although USGD yielded a much better objective function value than the SS and DS initialization methods on the 200-D Brown Almost-Linear problem, the performance of DYCORS-DDSRBF (USGD) is somewhat worse than DYCORS-DDSRBF (SS) and DYCORS-DDSRBF (DS) on this problem.

Figures 4–7 also show that PS-RBF (USGD) is better than both PS-RBF (SS) and PS-RBF (DS) on 12 of the 200-D test problems (all except Variably Dimensioned, Brown Almost-Linear, Discrete Boundary Value, and Linear Function–Full Rank). Moreover, the performance of PS-RBF (USGD) is comparable to that of the latter algorithms on the 200-D Variably Dimensioned and Discrete Boundary Value problems. However, the performance of PS-RBF (USGD) is somewhat worse than that of PS-RBF (SS) and PS-RBF (DS) on the 200-D Brown Almost-Linear in the later iterations, even though Table 2 shows that the former started with better

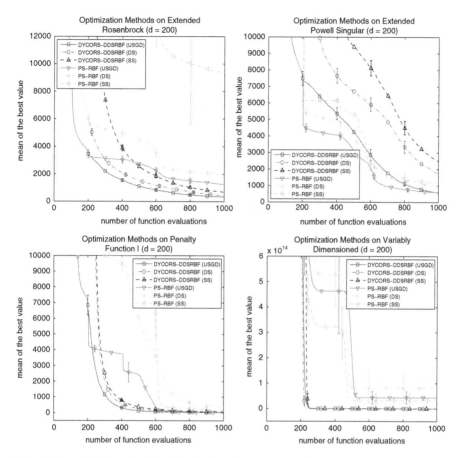

**Fig. 4** Mean of the best feasible objective function value (over 30 trials) vs number of function evaluations for the alternative optimization methods on the 200-D test problems. Error bars represent 95% t-confidence intervals for the mean

objective function values. In addition, PS-RBF (USGD) performed slightly worse than PS-RBF (DS) on the 200-D Linear Function–Full Rank even though the former also started with better objective function values.

Next, Figs. 8–10 (see Appendix) show that DYCORS-DDSRBF (USGD) is better than both DYCORS-DDSRBF (SS) and DYCORS-DDSRBF (DS) on 12 of the 1000-D test problems (all except Extended Powell Singular, Variably Dimensioned, Rastrigin, and Keane). Moreover, the performance of the former is comparable to that of the latter algorithms on the 1000-D Variably Dimensioned and Rastrigin problems.

Figures 8 and 10 also show that PS-RBF (USGD) is better than both PS-RBF (SS) and PS-RBF (DS) on seven of the 1000-D test problems (Extended Rosenbrock, Brown Almost-Linear, Broyden Banded, Linear Function–Full Rank, Ackley, Rastrigin, and Griewank). Moreover, its performance is comparable to those of the

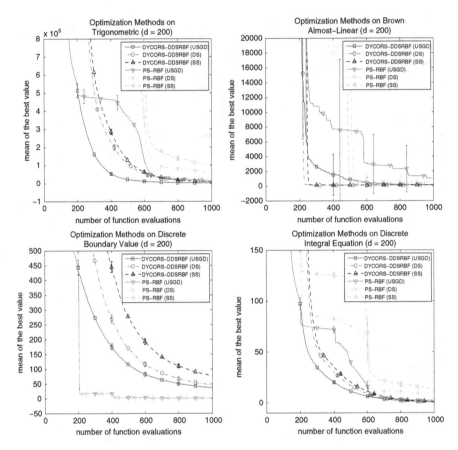

**Fig. 5** Mean of the best feasible objective function value (over 30 trials) vs number of function evaluations for the alternative optimization methods on the 200-D test problems. Error bars represent 95% t-confidence intervals for the mean

latter algorithms on six other 1000-D test problems (Extended Powell Singular, Penalty Function I, Variably Dimensioned, Discrete Boundary Value, Discrete Integral Equation, and Broyden Tridiagonal). However, PS-RBF (USGD) is worse than PS-RBF (DS) on the 1000-D Trigonometric, Linear Function–Rank 1, and Keane problems.

Overall, USGD yielded improvements for DYCORS-DDSRBF over the simpler SS and DS initialization on the 72-D groundwater application and on most of the 200-D and 1000-D test problems. USGD also yielded improvements for PS-RBF over SS and DS on GWB72 and on many of the 200-D problems. Moreover, it yielded improvements for PS-RBF on only seven of the 1000-D problems, but it did not really hurt performance on most of the remaining 1000-D test problems. As noted earlier, the main reason USGD was more effective for DYCORS-DDSRBF is that the USGD points are actually used by the algorithm, whereas only the best of the USGD points is used by PS-RBF.

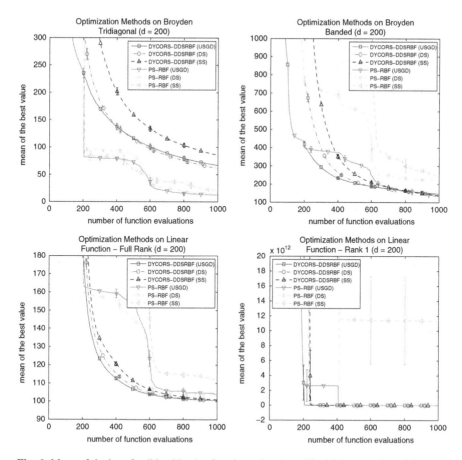

**Fig. 6** Mean of the best feasible objective function value (over 30 trials) vs number of function evaluations for the alternative optimization methods on the 200-D test problems. Error bars represent 95% t-confidence intervals for the mean

## 4.7 *Running Times*

To get an idea of the overhead computational effort required by the different algorithms, Table 6 reports the average running times on the 1000-D Extended Rosenbrock problem for 2000 function evaluations, excluding the time required by the initialization procedures. Note that DYCORS-DDSRBF has much longer average running times compared to PS-RBF. However, for truly expensive problems where each function evaluation could take hours, these running times are still much smaller than the total time required for all function evaluations.

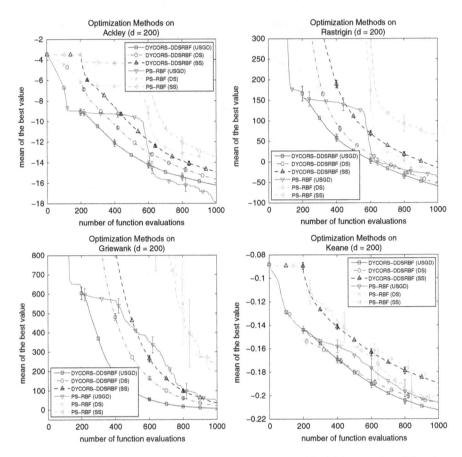

**Fig. 7** Mean of the best feasible objective function value (over 30 trials) vs number of function evaluations for the alternative optimization methods on the 200-D test problems with many local minima. Error bars represent 95% t-confidence intervals for the mean

## 5   Summary and Conclusions

This paper presented an initialization strategy for surrogate-based optimization method called underdetermined simplex gradient descent (USGD) that generates a set of $d + 1$ affinely independent points in $\mathbb{R}^d$ while making progress towards the optimum and also while keeping the condition number of the corresponding linear interpolation matrix within a reasonable value. Numerical experiments on 200-D and 1000-D instances of 16 well-known test problems and on a 72-D groundwater bioremediation application suggest that this approach results in much better objective function values compared to simpler and more standard initialization strategies, called *static simplex (SS)* and *dynamic simplex (DS)*.

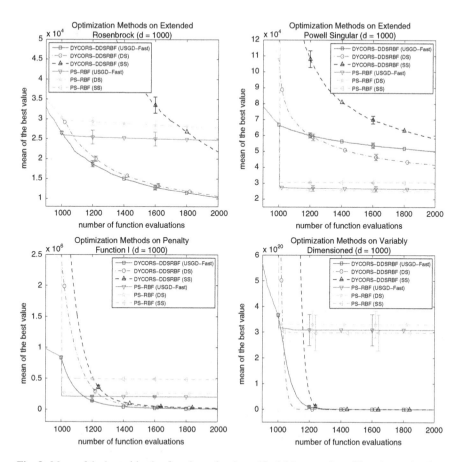

**Fig. 8** Mean of the best objective function value (over 10 trials) vs number of function evaluations for the alternative optimization methods on the 1000-D test problems. Error bars represent 95% t-confidence intervals for the mean

This paper also compared the performance of the DYCORS-DDSRBF algorithm [56] initialized by USGD with the same algorithm initialized by the SS and DS strategies on the same test problems. Moreover, a pattern search algorithm that uses RBF surrogates (PS-RBF) [38] initialized by USGD was also compared with the same algorithm initialized by the SS and DS procedures. The numerical results given in the performance and data profiles and also the average progress curves consistently showed that DYCORS-DDSRBF (USGD) was a substantial improvement over both DYCORS-DDSRBF (SS) and DYCORS-DDSRBF (DS) on the 72-D groundwater bioremediation problem and on the 200-D and 1000-D test problems. Similarly, PS-RBF (USGD) was also generally an improvement over both PS-RBF (SS) and PS-RBF (DS) on the same set of problems. Overall, the numerical results suggest that the USGD initialization strategy is promising

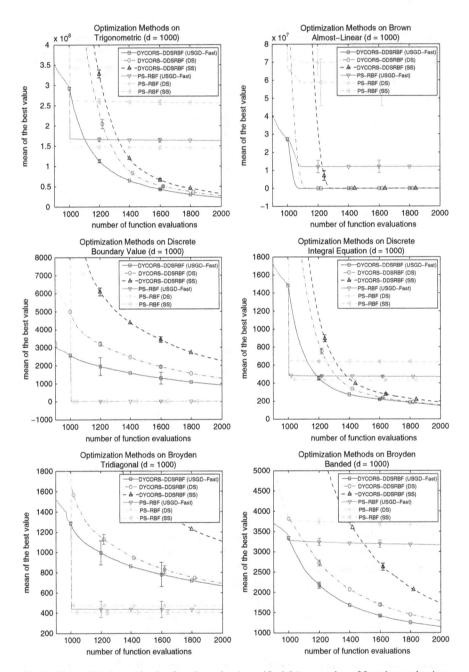

**Fig. 9** Mean of the best objective function value (over 10 trials) vs number of function evaluations for the alternative optimization methods on the 1000-D test problems. Error bars represent 95% t-confidence intervals for the mean

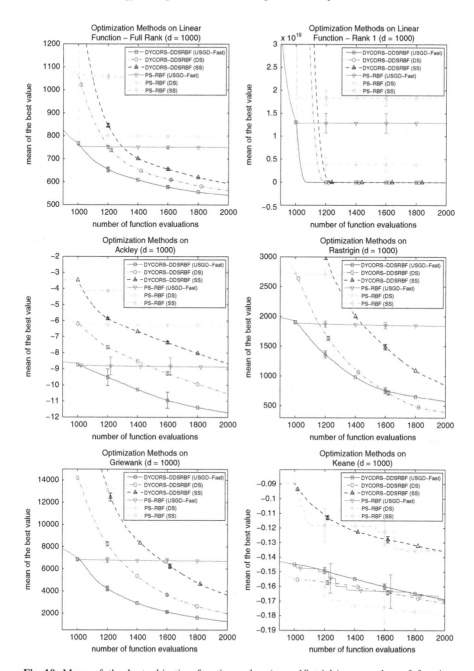

**Fig. 10** Mean of the best objective function value (over 10 trials) vs number of function evaluations for the alternative optimization methods on the 1000-D test problems. Error bars represent 95% t-confidence intervals for the mean

**Table 6** Average running times (over 10 trials) of the surrogate-based methods for 2000 function evaluations on the 1000-D Extended Rosenbrock problem

| Algorithm | Average running time |
|---|---|
| DYCORS-DDSRBF (USGD-Fast) | 6070.18 s (1.69 h) |
| DYCORS-DDSRBF (DS) | 6146.93 s (1.71 h) |
| DYCORS-DDSRBF (SS) | 6127.76 s (1.70 h) |
| PS-RBF (USGD-Fast) | 370.34 s |
| PS-RBF (DS) | 390.42 s |
| PS-RBF (SS) | 475.85 s |

These running times exclude the times required by the initialization procedures

for surrogate-based algorithms for very high-dimensional expensive black-box problems involving hundreds of decision variables when only a relatively limited number of function evaluations can be performed.

**Acknowledgements** I would like to thank Ismael Vaz and Luís Vicente for the PSwarm package, that includes a pattern search Matlab code with options for combining with RBF models. I am also grateful to Jorge Moré and Stefan Wild for their Matlab code that creates performance and data profiles.

# Appendix

Figures 8–10 show the average progress curves for the surrogate-based optimization algorithms on the 1000-D test problems. Because the computational overheads of running some of these methods are enormous, the algorithms are only run for 10 trials instead of 30 trials on each problem.

# References

1. Abramson, M.A., Audet, C.: Convergence of mesh adaptive direct search to second-order stationary points. SIAM J. Optim. **17**(2), 606–619 (2006)
2. Aleman, D.M., Romeijn, H.E., Dempsey, J.F.: A response surface approach to beam orientation optimization in intensity modulated radiation therapy treatment planning. INFORMS J. Comput. **21**(1), 62–76 (2009)
3. Audet, C., Dennis, J.E., Jr.: Mesh adaptive direct search algorithms for constrained optimization. SIAM J. Optim. **17**(2), 188–217 (2006)
4. Audet, C., Dennis, J.E., Jr.: Le Digabel S.: Parallel space decomposition of the mesh adaptive direct search algorithm. SIAM J. Optim. **19**(3), 1150–1170 (2008)
5. Bettonvil, B., Kleijnen J.P.C.: Searching for important factors in simulation models with many factors: Sequential bifurcation. Eur. J. Oper. Res. **96**(1), 180–194 (1997)

6. Björkman, M., Holmström, K.: Global optimization of costly nonconvex functions using radial basis functions. Optim. Eng. **1**(4), 373–397 (2000)
7. Booker, A.J., Dennis, J.E., Jr., Frank, P.D., Serafini, D.B., Torczon, V., Trosset M.W.: A rigorous framework for optimization of expensive functions by surrogates. Struct. Optim. **17**(1), 1–13 (1999)
8. Buhmann, M.D.: Radial Basis Functions. Cambridge University Press, Cambridge (2003)
9. Bull, A.D.: Convergence rates of efficient global optimization algorithms. J. Mach. Learn. Res. **12**(Oct), 2879–2904 (2011)
10. Cassioli, A., Schoen, F.: Global optimization of expensive black box problems with a known lower bound. J. Global Optim. (2011). doi: 10.1007/s10898-011-9834-7
11. Chambers, M., Mount-Campbell, C.A.: Process optimization via neural network metamodeling. Int. J. Prod. Econ. **79**(2), 93–100 (2002)
12. Chen, L.-L., Liao, C., Lin, W.-B., Chang, L., Zhong, X.-M.: Hybrid-surrogate-model-based efficient global optimization for high-dimensional antenna design. Prog. Electromagnetics Res. **124**, 85–100 (2012)
13. Conn, A.R., Scheinberg, K., Vicente, L.N.: Geometry of interpolation sets in derivative free optimization. Math. Program. **111**(1–2), 141–172 (2008a)
14. Conn, A.R., Scheinberg, K., Vicente, L.N.: Geometry of sample sets in derivative-free optimization: polynomial regression and underdetermined interpolation. IMA J. Numer. Anal. **28**(4), 721–748 (2008b)
15. Conn, A.R., Scheinberg, K., Vicente, L.N.: Global convergence of general derivative-free trust-region algorithms to first- and second-order critical points. SIAM J. Optim. **20**(1), 387–415 (2009a)
16. Conn, A.R., Scheinberg, K., Vicente, L.N.: Introduction to Derivative-Free Optimization. SIAM, Philadelphia, PA (2009b)
17. Conn, A.R., Scheinberg, K., Toint, Ph.L.: Recent progress in unconstrained nonlinear optimization without derivatives. Math. Program. **79**(3), 397–414 (1997)
18. Conn, A.R., Le Digabel, S.: Use of quadratic models with mesh-adaptive direct search for constrained black box optimization. Optim. Meth. Software **28**(1), 139–158 (2013)
19. Cressie, N.: Statistics for Spatial Data. Wiley, New York (1993)
20. Custódio, A.L., Rocha, H., Vicente, L.N.: Incorporating minimum Frobenius norm models in direct search. Computat. Optim. Appl. **46**(2), 265–278 (2010)
21. Custódio, A.L., Vicente, L.N.: Using sampling and simplex derivatives in pattern search methods. SIAM J. Optim. **18**(2), 537–555 (2007)
22. Egea, J.A., Vazquez, E., Banga, J.R., Marti, R.: Improved scatter search for the global optimization of computationally expensive dynamic models. J. Global Optim. **43**(2–3), 175–190 (2009)
23. García-Palomares, U.M., García-Urrea, I.J., Rodríguez-Hernández, P.S.: On sequential and parallel non-monotone derivative-free algorithms for box constrained optimization. Optim. Meth. Software (2012). doi:10.1080/10556788.2012.693926
24. Gray, G.A., Kolda, T.G.: Algorithm 856: APPSPACK 4.0: asynchronous parallel pattern search for derivative-free optimization. ACM Trans. Math. Software **32**(3), 485–507 (2006)
25. Gutmann, H.-M.: A radial basis function method for global optimization. J. Global Optim. **19**(3), 201–227 (2001)
26. Hansen, N.: The CMA evolution strategy: a comparing review. In: Lozano, J.A., Larranga, P., Inza, I., Bengoetxea, E. (eds.) Towards a New Evolutionary Computation, pp. 75–102, Springer, Berlin (2006)
27. Hansen, N., Ostermeier, A.: Completely derandomized self-adaptation in evolution strategies. Evol. Comput. **9**(2), 159–195 (2001)
28. Holmström, K.: An adaptive radial basis algorithm (ARBF) for expensive black-box global optimization. J. Global Optim. **41**(3), 447–464 (2008)
29. Huang, D., Allen, T.T., Notz, W.I., Zeng, N.: Global optimization of stochastic black-box systems via sequential kriging meta-models. J. Global Optim. **34**(3), 441–466 (2006)

30. Jakobsson, S., Patriksson, M., Rudholm, J., Wojciechowski, A.: A method for simulation based optimization using radial basis functions. Optim. Eng. **11**(4), 501–532 (2010)
31. Jin Y.: Surrogate-assisted evolutionary computation: recent advances and future challenges. Swarm Evol. Comput. **1**(2), 61–70 (2011)
32. Jin, Y., Olhofer, M., Sendhoff, B.: A framework for evolutionary optimization with approximate fitness functions. IEEE Trans. Evol. Comput. **6**(5), 481–494 (2002)
33. Jones, D.R., Schonlau, M., Welch, W.J.: Efficient global optimization of expensive black-box functions. J. Global Optim. **13**(4), 455–492
34. Jones, D.R.: Large-scale multi-disciplinary mass optimization in the auto industry. Presented at the Modeling and Optimization: Theory and Applications (MOPTA) 2008 Conference, Ontario, Canada (2008)
35. Kolda, T.G., Lewis, R.M., Torczon, V.: Optimization by direct search: new perspectives on some classical and modern methods. SIAM Rev. **45**(3), 385–482 (2003)
36. Kolda, T.G., Torczon, V.J.: On the convergence of asynchronous parallel pattern search. SIAM J. Optim. **14**(4), 939–964 (2004)
37. Le Digabel, S.: Algorithm 909: NOMAD: Nonlinear optimization with the MADS algorithm. ACM Trans. Math. Software **37**(4), 44:1–44:15 (2011)
38. Le Thi, H.A., Vaz, A.I.F., Vicente, L.N.: Optimizing radial basis functions by D.C. programming and its use in direct search for global derivative-free optimization. TOP **20**(1), 190–214 (2012)
39. Loshchilov, I., Schoenauer, M., Sebag, M.: Self-adaptive surrogate-assisted covariance matrix adaptation evolution strategy. In: Proceedings of the Genetic and Evolutionary Computation Conference (GECCO-2012), ACM Press, New York (2012)
40. Marsden, A.L., Wang, M., Dennis, J.E., Jr., Moin, P.: Optimal aeroacoustic shape design using the surrogate management framework. Optim. Eng. **5**(2), 235–262 (2004)
41. Minsker, B.S., Shoemaker, C.A.: Dynamic optimal control of in-situ bioremediation of groundwater. J. Water Resour. Plann. Manag. **124**(3), 149–161 (1998)
42. Moré, J., Wild, S.: Benchmarking derivative-free optimization algorithms. SIAM J. Optim. **20**(1), 172–191 (2009)
43. Moré, J., Garbow, B., Hillstrom, K.: Testing unconstrained optimization software. ACM Trans. Math. Software **7**(1), 17–41 (1981)
44. Myers, R.H., Montgomery, D.C.: Response Surface Methodology: Process and Product Optimization Using Designed Experiments, 3rd edn. Wiley, New York (2009)
45. Oeuvray, R., Bierlaire, M.: BOOSTERS: A derivative-free algorithm based on radial basis functions. Int. J. Model. Simulat. **29**(1), 26–36 (2009)
46. Oeuvray, R.: Trust-region methods based on radial basis functions with application to biomedical imaging. *Ph.D. thesis*, École Polytechnique Fédérale de Lausanne (EPFL), Lausanne, Switzerland (2005)
47. Parr, J.M., Keane, A.J., Forrester, A.I.J., Holden, C.M.E.: Infill sampling criteria for surrogate-based optimization with constraint handling. Eng. Optim. **44**(10), 1147–1166 (2012)
48. Plantenga, T., Kolda, T.: HOPSPACK: Software framework for parallel derivative-free optimization. Sandia Technical Report (SAND 2009–6265). (2009)
49. Powell, M.J.D.: The theory of radial basis function approximation in 1990. In: Light, W. (ed.) Advances in Numerical Analysis, Volume 2: Wavelets, Subdivision Algorithms and Radial Basis Functions. pp. 105–210. Oxford University Press, Oxford (1992)
50. Powell, M.J.D.: UOBYQA: Unconstrained optimization by quadratic approximation. Math. Program. **92**(3), 555–582 (2002)
51. Powell, M.J.D.: The NEWUOA software for unconstrained optimization without derivatives. In: Di Pillo, G., Roma, M. (eds.) Large-Scale Nonlinear Optimization, pp. 255–297. Springer, New York (2006)
52. Regis, R.G.: Stochastic radial basis function algorithms for large-scale optimization involving expensive black-box objective and constraint functions. Comput. Oper. Res. **38**(5), 837–853 (2011)

53. Regis, R.G.: Constrained optimization by radial basis function interpolation for high-dimensional expensive black-box problems with infeasible initial points. Eng. Optim. (2013). doi: 10.1080/0305215X.2013.765000.
54. Regis, R.G., Shoemaker, C.A.: A stochastic radial basis function method for the global optimization of expensive functions. INFORMS J. Comput. **19**(4), 497–509 (2007a)
55. Regis, R.G., Shoemaker, C.A.: Improved strategies for radial basis function methods for global optimization. J. Global Optim. **37**(1), 113–135 (2007b)
56. Regis, R.G., Shoemaker, C.A.: A quasi-multistart framework for global optimization of expensive functions using response surface models. J. Global Optim. (2012). doi: 10.1007/s10898-012-9940-1
57. Regis, R.G., Shoemaker, C.A.: Combining radial basis function surrogates and dynamic coordinate search in high-dimensional expensive black-box optimization. Eng. Optim. **45**(5), 529–555 (2013)
58. Rocha, H., Dias, J.M., Ferreira, B.C., Lopes, M.C.: Selection of intensity modulated radiation therapy treatment beam directions using radial basis functions within a pattern search methods framework. J. Global Optim. (2012). doi: 10.1007/s10898-012-0002-5
59. Sacks, J., Welch, W.J., Mitchell, T.J., Wynn, H.P.: Design and analysis of computer experiments. Stat. Sci. **4**(4), 409–435 (1989)
60. Scheinberg, K., Toint, Ph.L.: Self-correcting geometry in model-based algorithms for derivative-free unconstrained optimization. SIAM J. Optim. **20**(6), 3512–3532 (2010)
61. Shan, S., Wang, G.: Survey of modeling and optimization strategies to solve high-dimensional design problems with computationally-expensive black-box functions. Struct. Multidisciplinary Optim. **41**(2), 219–241 (2010)
62. Shan, S., Wang, G.G.: Metamodeling for high dimensional simulation-based design problems. ASME Journal of Mechanical Design, **132**(5), 051009 (2011)
63. Shoemaker, C.A., Willis, M., Zhang, W., Gossett, J.: Model analysis of reductive dechlorination with data from Cape Canaveral field site. In: Magar, V., Vogel, T., Aelion, C., Leeson, A. (eds.) Innovative Methods in Support of Bioremediation, pp. 125–131. Battelle Press, Columbus, OH (2001)
64. Tolson, B.A., Shoemaker, C.A.: Dynamically dimensioned search algorithm for computationally efficient watershed model calibration. Water Resour. Res. **43**, W01413 (2007) doi:10.1029/2005WR004723.
65. Torczon, V.: On the convergence of pattern search algorithms. SIAM J. Optim. **7**(1), 1–25 (1997)
66. Vapnik, V.: The Nature of Statistical Learning Theory. Springer, New York (1995)
67. Vaz, A.I.F., Vicente, L.N.: A particle swarm pattern search method for bound constrained global optimization. J. Global Optim. **39**(2), 197–219 (2007)
68. Vaz, A.I.F., Vicente, L.N.: PSwarm: A hybrid solver for linearly constrained global derivative-free optimization. Optim. Meth. Software **24**(4–5), 669–685 (2009)
69. Vazquez, E., Bect, J.: Convergence properties of the expected improvement algorithm with fixed mean and covariance functions. J. Stat. Plann. Infer. **140**(11), 3088–3095 (2010)
70. Viana, F.A.C., Haftka, R.T., Watson, L.T.: Why not run the efficient global optimization algorithm with multiple surrogates?. 51th AIAA/ASME /ASCE/AHS/ASC Structures, Structural Dynamics, and Materials Conference, AIAA 2010–3090, Orlando (2010)
71. Villemonteix, J., Vazquez, E., Walter, E.: An informational approach to the global optimization of expensive-to-evaluate functions. J. Global Optim. **44**(4), 509–534 (2009)
72. Wild, S.M., Regis, R.G., Shoemaker, C.A.: ORBIT: Optimization by radial basis function interpolation in trust-regions. SIAM J. Sci. Comput. **30**(6), 3197–3219 (2008)
73. Wild, S.M., Shoemaker, C.A.: Global convergence of radial basis function trust region derivative-free algorithms. SIAM J. Optim. **21**(3), 761–781 (2011)
74. Yoon, J.-H., Shoemaker, C.A.: Comparison of optimization methods for ground-water bioremediation. J. Water Resour. Plann. Manag. **125**(1), 54–63 (1999)
75. The MathWorks, Inc. Matlab Optimization Toolbox: User's Guide, Version 4. Natick, MA (2009)

# Smoothing and Regularization for Mixed-Integer Second-Order Cone Programming with Applications in Portfolio Optimization

**Hande Y. Benson and Ümit Sağlam**

**Abstract** Second-order cone programming problems (SOCPs) have been well studied in literature, and computationally efficient implementations of solution algorithms exist. In this paper, we study an extension: mixed-integer second-order cone programming problems (MISOCPs). Our focus is on designing an algorithm for solving the underlying SOCPs as nonlinear programming problems (NLPs) within two existing frameworks, branch-and-bound and outer approximation, for mixed-integer nonlinear programming (MINLP). We pay particular attention to resolving the nondifferentiability of the underlying SOCPs via a smoothing reformulation, as well as warmstarting and infeasibility detection using a regularization. We investigate the application of our proposed techniques to portfolio optimization problems that can be formulated as MISOCPs, and preliminary numerical results using the Matlab-based optimization package MILANO (Benson, MILANO—a Matlab-based code for mixed-integer linear and nonlinear optimization. http://www.pages.drexel.edu/~hvb22/milano) are provided.

**Keywords** Mixed-integer second-order cone programming • Portfolio optimization • Second-order cone programming

## 1 Introduction

In this paper, we study mixed-integer second-order cone programming problems (MISOCPs) of the form

$$
\begin{aligned}
&\min_{x \in \mathcal{X}} \; c^T x \\
&\text{s.t.} \;\; \|A_i x + b_i\| \le a_{0i}^T x + b_{0i}, \qquad i = 1, \dots, m,
\end{aligned}
\tag{1}
$$

H.Y. Benson (✉) • Ü. Sağlam
Drexel University, Philadelphia, PA, USA
e-mail: benson@drexel.edu; us26@drexel.edu

L.F. Zuluaga and T. Terlaky (eds.), *Modeling and Optimization: Theory and Applications*,
Springer Proceedings in Mathematics & Statistics 62, DOI 10.1007/978-1-4614-8987-0_4,
© Springer Science+Business Media New York 2013

where $x$ is the $n$-vector of decision variables, $\mathcal{X} = \{(y,z) : y \in \mathcal{Z}^p, z \in \mathcal{R}^k, p + k = n\}$, and the data are $c \in \mathcal{R}^n$, $A_i \in \mathcal{R}^{m_i \times n}$, $b_i \in \mathcal{R}^{m_i}$, $a_{0i} \in \mathcal{R}^n$, and $b_{0i} \in \mathcal{R}$ for $i = 1,\ldots,m$. The notation $\| \cdot \|$ denotes the Euclidean norm, and the constraints are said to define the *second-order cone*, also referred to as the *Lorentz cone*.

When $p = 0$, (1) reduces to a second-order cone programming problem (SOCP) in the so-called dual form. SOCPs have been widely studied in literature; see [43] for a general introduction and an extensive list of applications and [1] for an overview of the general properties, duality theory, and interior-point methods for this class of problems. In fact, interior-point methods have been the most popular solution methods for this class of problems, enjoying both good theoretical convergence properties [49] and efficient computational performance in implementations such as SEDUMI [57], SDPT3 [59], MOSEK [2], and CPLEX [35]. Besides interior-point methods designed specifically for SOCPs, there is also a sequential linear programming method [39], a simplex method for conic programs [33], an interior-point method for nonlinear programming that treats SOCPs as a special case [16], as well as approaches based on polyhedral reformulations of the second-order cone [8].

Comparatively, MISOCP is a less mature field, but these problems arise in a variety of application areas, and there have been significant advances in solution algorithms in the last decade. We present a literature review covering these algorithms and many application areas in the next section. In Sect. 3, we propose and implement two approaches for MISOCP, based on popular algorithms for mixed-integer nonlinear programming (MINLP): a branch-and-bound method and an outer approximation method. Both algorithms use a version of the primal–dual penalty interior-point method proposed in [13, 14] for solving the underlying SOCPs, which allows us to perform warmstarts and detect infeasibilities in an efficient manner. In addition, we reformulate the second-order cone constraint as discussed in [16] in order to convert the underlying SOCPs into smooth convex nonlinear programming problems (NLP). Our main purpose in this paper is to solve certain types of portfolio optimization problems which are formulated as MISOCPs, and we discuss these problems and their formulation in depth in Sect. 4. Numerical testing for these portfolio optimization problems has been conducted using the Matlab-based solver MILANO [9] and is documented in Sect. 5.

## 2 Literature Review

In this section, we provide a thorough overview of the existing literature on MISOCPs. While the field may not be as mature as SOCP, there have been a wide variety of algorithms studied in the last decade, and the application areas range from supply chain management to electrical engineering to asset pricing. We will focus on portfolio optimization models in Sect. 4 and will provide a literature review for this particular application area then.

## 2.1 Algorithms for Solving MISOCPs

One very straightforward way to devise a method for solving MISOCPs is to use a branch-and-bound algorithm that calls an interior-point method designed specifically for SOCPs at each node. However, if such a method is to be competitive on large-scale MISOCPs, it is important to reduce the number of nodes in the tree using valid inequalities and cuts designed specifically for MISOCP and to reduce the runtime at each node using an SOCP solver that is capable of warmstarting and infeasibility detection.

There are other approaches for MINLP besides branch-and-bound which can similarly be adopted for the case of MISOCP, using the fact that the underlying SOCPs are essentially convex NLPs. These approaches include outer approximation [29], extended cutting-plane methods [61], and generalized Benders' decomposition [32]. However, the nondifferentiability of the constraint functions in (1) is of particular concern when generating the gradient-based cuts required by these methods, and their application to MISOCP should be done carefully and by considering this special case.

Additionally, any method that can convert the underlying SOCPs into linear programming problems can take advantage of the efficient algorithms designed for mixed-integer linear programming.

A number of studies appear in literature dealing with algorithms specifically for MISOCP:

- In [21], Cezik and Iyengar study 0–1 mixed-integer conic programming problems (MICPs), a set of problems to which many mixed-integer linear programming problems and MISOCPs belong either directly or via Dantzig–Wolfe decomposition. Their approach is to extend some well-known techniques for mixed-integer linear programming to mixed-integer programs involving second-order cone and/or semidefinite constraints. They show that Gomory cuts can be readily extended to the conic case and that all valid inequalities for the convex hull of the feasible region of a 0–1 MICP can be generated using their extension of the Chvátal–Gomory procedure a finite number of times. In addition, they propose relaxations that are tighter than the continuous relaxation of an MICP by extending approaches of [6,41,44,53,54], all originally developed for integer or mixed-integer linear programming. The Gomory cuts and the tight relaxations are used in a cut algorithm for MICP. The authors note, however, that their proposed approach suffers due to their use of an interior-point method to solve the underlying conic programs, both in terms of the lack of depth of the proposed cuts at extreme points and of the lack of efficiency due to lack of warmstarting strategies. They present preliminary numerical results and pointers for future improvement.

- In [4], Atamturk and Narayanan focus on MISOCPs and their solution using a branch-and-bound framework. They introduce mixed-integer rounding cuts that are added at the root node of the branch-and-bound tree, and their preliminary results show that the cuts can significantly reduce the number of nodes in the tree. They provide a more thorough analysis and further examples showing the success of their approach in [5].

- In [60], Vielma et al. propose using a lifted polyhedral relaxation [8] of the underlying SOCPs, thereby solving the MISOCPs using a linear programming-based branch-and-bound framework. Their approach can be generalized to any convex MINLP, does not use gradients to generate the cuts, and benefits from the linear programming structure that can use a simplex-based method with warmstarting capabilities within the solution process. Numerical studies on portfolio optimization problems show that the method outperforms CPLEX and Bonmin. A similar method is used in [55] by Soberanis to solve the MISOCP reformulations of risk optimization problems with $p$-order conic constraints.
- In [26], Drewes proposes both a branch-and-cut method and a hybrid branch-and-bound/outer approximation method for solving MISOCPs. The branch-and-cut method uses techniques similar to [21] and those developed in [56] for mixed-integer convex optimization problems with binary variables. The hybrid approach extends outer approximation methods, which use gradient-based techniques to generate cuts, to the case of MISOCPs using subgradients. Infeasibility detection for branch-and-bound method is handled using an auxiliary, infeasibility-minimizing problem, and the creation of strictly feasible interior points for the branch-and-cut method is accomplished through a penalty approach. Numerical results on a number of test problems are provided.
- In [7], Belotti et al. propose disjunctive conic and cylindrical cuts for mixed-integer conic programming and provide conditions for such cuts to exist. They show that such a cone and cylinder exist for the case of MISOCP and provide a method for finding the cuts, to be included in a branch-and-cut framework for MISOCP. They provide several numerical examples to illustrate these cuts and compare their approach to [4]. In addition, the split disjunctions discussed in [23] by Dadush et al. use the same approach at [7].

In this paper, we will propose two algorithms, one based on a branch-and-bound framework and the other on outer approximation and both using the same primal–dual penalty interior-point method to solve the underlying SOCPs. We will address issues of warmstarting, infeasibility detection, and nondifferentiability in the design of our algorithms. Any number of cuts discussed in the literature can be incorporated into these approaches for further speedup.

## 2.2 MISOCPs Arising in Applications

In Sect. 4, we will discuss portfolio optimization problems with second-order cone constraints and binary variables. There are a variety of other application areas in which the problems can be formulated as MISOCPs, and we list several of them here:

- In [51], Pinar describes a pricing problem for an American option in a financial market under uncertainty. The multiperiod, discrete time, and discrete state space structure is modeled using a scenario tree, and, therefore, the resulting problem is large scale. The second-order cone constraints arise as risk constraints that provide a lower bound for the Sharpe ratio of the final wealth position of the buyer. The binary variables are introduced to denote the decision whether to exercise the option at each node of the scenario tree, and additional constraints enforce that the option is exercised at no more than 1 node in each sample path. The numerical results show that these problems have over 20,000 continuous variables, 5,000 discrete variables, and 30,000 constraints and are quite challenging for existing MISOCP software.

- In [34], Hijazi et al. investigate a telecommunications network problem that seeks to minimize the network response time to a user request. The binary variables correspond to paths that can be used to route part or all of a request from each network commodity to each user, and each one denotes whether or not a path is open. The delay along each open path is uncertain and is modeled using a robust constraint. These constraints are also disjunctive since they are only used if the path is open, and a perspective function approach is used to handle the disjunctivity, leading to their reformulation as second-order cone constraints in the resulting model. The numerical testing shows that CPLEX has trouble solving large MISOCP instances, while related MINLPs are solved within reasonable time requirements by Bonmin.

- In [22], Cheng et al. model and solve a coordinated multipoint transmission problem for cellular networks. The binary variables represent the assignment of mobile units to base stations, where multiple base stations can coordinate the transmission. The second-order cone constraints are formulated for each base station and serve to limit the total power transmitted from the base station to all of the mobile units that it serves. Due to the large size of the problem, the authors propose a heuristic, which is able to obtain slightly worse solutions in significantly less CPU times than CPLEX.

- In [19], Brandenberg and Roth introduce a new algorithm for the Euclidean $k$-center problem, which deals with the clustering of a group of points among $k$ balls and arises in facility location and data classification applications. The binary variables denote the assignment of the points to the balls, and second-order cone constraints are used to denote that if a point is assigned to a ball, then the Euclidean distance between the point and the center of the ball must be no more than the radius of the ball. Numerical results are obtained using a branch-and-bound code calling SeDuMi.

- In [45], Mak et al. consider the problem of creating a network infrastructure and providing coverage for battery-swapping stations to service electric vehicles. Given an existing freeway network, they consider candidate locations and use a binary variable for each candidate to denote whether or not a swapping station is located there. Additional binary variables denote whether vehicles traveling along portions of the network will visit a swapping station. Since there is uncertainty in the number of electric vehicles that travel along the paths throughout the network, the authors introduce a second-order cone constraint, similar to the conditional value-at-risk constraint we will use later in this paper, as a robust service-level requirement. They solve the problem with data from the San Francisco freeway network using CPLEX.

- In [27], Du et al. present an MISOCP as a relaxation of the MINLP that arises in the problem of determining the berthing positions and order for a group of vessels waiting at a container terminal in order to minimize the total waiting time of the vessels. Binary variables are used to denote the relative positions of pairs of vessels (whether one vessel is to the left of another and whether one vessel is earlier than another). The second-order cone constraints arise in a reformulation of a nonlinear fuel consumption constraint. Numerical results are conducted for instances up to 28 vessels using CPLEX, which has runtime and memory problems as the problem size grows.

- In [3], Atamturk et al. explore a joint facility location and inventory management model under stochastic retailer demand. The binary variables arise in the choice of candidate locations at which to open distribution centers and the assignment of retailers to the distribution centers. The second-order cone constraints appear in the reformulation of the uncapacitated problem to move the nonlinear objective function terms denoting the fixed costs of placing and shipping orders and the expected safety stock cost into the constraints. The model with capacities additionally has a second-order cone constraint arising from moving an objective function term for the average inventory holding cost into a constraint and another one arising from the reformulation of a capacity constraint. Other related models with similar features are provided in the paper. Numerical results are conducted using algorithms studied in [50, 52] and CPLEX.

- In [58], Taylor and Hover present several problems from power distribution system reconfiguration, one of which is formulated as an MISOCP. The second-order cone constraints arise in the approximation of flow distribution equations. The binary variables appear as switching variables. Numerical results are presented on 32 to 880 bus systems using CPLEX.

## 3   Two Algorithms for Solving MISOCPs

In this section, we will describe two MINLP approaches that we have adapted for MISOCP. As stated, there are three important issues to consider: nondifferentiability of the underlying SOCP, warmstarting when solving a sequence of SOCPs, and infeasibility detection. We will address the first using a smooth convex reformulation of the SOCP and the latter two using a primal–dual penalty interior-point method.

## 3.1 The Ratio Reformulation

In [16], Benson and Vanderbei investigated the nondifferentiability of an SOCP and proposed several reformulations of the second-order cone constraint to overcome this issue. Note that the nondifferentiability is only an issue if it occurs at the optimal solution. Since an initial solution can be randomized, especially when using an infeasible interior-point method to solve the SOCP, the probability of encountering a point of nondifferentiability is 0.

For a constraint of the form

$$\|u\| \le t, \tag{2}$$

where $u$ is a vector and $t$ is a scalar, Benson and Vanderbei proposed the following:

- Exponential reformulation: Replacing (2) with $e^{(u^T u - t^2)/2} \le 1$ and $t \ge 0$ gives a smooth and convex reformulation of the problem, but numerical issues frequently arise due to the exponential.
- Smoothing by perturbation: Introducing a scalar variable $v$ into the norm gives a constraint of the form $\sqrt{v^2 + u^T u} \le t$, but in order for the formulation to be smooth, we need $v > 0$. This is ensured by setting $v \ge \epsilon$ for a small constant $\epsilon$, usually taken around $10^{-6} - 10^{-4}$.
- Ratio reformulation: Replacing (2) with $\frac{u^T u}{t} \le t$ and $t \ge 0$ yields a convex reformulation of the problem, but the constraint function may still not be smooth. Nevertheless, in many applications, such as the portfolio optimization problems to be studied in the next section, the right-hand side of the second-order cone constraint in (1) is either a scalar or bounded away from zero at the optimal solution.

While the exponential reformulation and smoothing by perturbation resolve the nondifferentiability issue for the general SOCP, the ratio reformulation will be our pick for this paper since we will focus only on portfolio optimization problems. In the case of the exponential reformulation, numerical difficulties were encountered during our testing, even for modest values of $(u^T u - t^2)/2$. For the perturbation approach, we found that the values of $u^T u$ were sufficiently from zero to alleviate any potential concerns with the ratio reformulation.

Applying the ratio reformulation, we will be solving the following MISOCP instead of (1)

$$
\begin{aligned}
\min_{x \in \mathcal{X}} \quad & c^T x \\
\text{s.t.} \quad & \frac{(A_i x + b_i)^T (A_i x + b_i)}{a_{0i}^T x + b_{0i}} \le a_{0i}^T x + b_{0i}, \quad i = 1, \ldots, m, \\
& a_{0i}^T x + b_{0i} \ge 0, \quad i = 1, \ldots, m.
\end{aligned} \tag{3}
$$

## 3.2 The Primal and Dual Penalty Problems

In order to solve the SOCPs that will arise during the course of the branch-and-bound and the outer approximation methods, we will use the primal–dual penalty interior-point method that was introduced in [13] for linear programming and in [14] for nonlinear programming. This approach includes relaxation/penalty terms in both the primal and the dual problems, which imbues the algorithm with the ability to perform warmstarts and detect infeasibilities. The new terms do not change the structure of the problem, that is, we will still solve an SOCP and can continue to use a highly efficient interior-point method to do so. In addition, the relaxation scheme creates strict interiors for the feasible regions of both the primal and the dual problems, thereby providing a regularization and allowing for the solution of SOCPs that may not otherwise satisfy standard assumptions for the interior-point method to work.

Even though our approach is to solve the SOCP as a nonlinear programming problem, the particular relaxation/penalty scheme differs slightly from the one presented in [14]. If we were to follow the outline of the approach presented in that paper, the relaxed SOCP constraint would have the form

$$
\begin{aligned}
\frac{(A_i x + b_i)^T (A_i x + b_i)}{a_{0i}^T x + b_{0i}} &\leq a_{0i}^T x + b_{0i} + \xi_i, \\
a_{0i}^T x + b_{0i} + \rho_i &\geq 0, \\
\xi_i, \rho_i &\geq 0,
\end{aligned}
$$

where $\xi, \rho \in \mathcal{R}^m$ are the relaxation variables that would get penalized in the objective function. While this would provide a sufficient relaxation for our purposes, we have decided to use the following relaxation instead:

$$
\begin{aligned}
\frac{(A_i x + b_i)^T (A_i x + b_i)}{a_{0i}^T x + b_{0i} + \xi_i} &\leq a_{0i}^T x + b_{0i} + \xi_i, \\
a_{0i}^T x + b_{0i} + \xi_i &\geq 0, \\
\xi_i &\geq 0.
\end{aligned}
\tag{4}
$$

This form of the relaxation can be obtained in two different ways:

- If we apply the relaxation scheme from [14] to the second-order cone constraint in (1), we obtain

$$
\|A_i x + b_i\| \leq a_{0i}^T x + b_{0i} + \xi_i, \qquad \xi_i \geq 0.
$$

Note that we have a second-order cone and a linear constraint after the relaxation. If we apply the ratio reformulation now, we obtain (4).
- The ratio reformulation constraint in (3) can also be written as a semidefinite constraint of the form

$$\begin{bmatrix} (a_{0i}^T x + b_{0i})I & A_i x + b_i \\ (A_i x + b_i)^T & a_{0i}^T x + b_{0i} \end{bmatrix} \succeq 0,$$

where $I$ is the $m_i \times m_i$ identity matrix. As outlined in [16], the semidefinite constraint is equivalent to the entries of the diagonal matrix $D$ in the $LDL^T$ factorization of the above matrix being nonnegative. Without permutation, we have that

$$D_{jj} = \begin{cases} a_{0i}^T x + b_{0i}, & j = 1 \ldots m_i, \\ a_{0i}^T x + b_{0i} - \dfrac{(A_i x + b_i)^T (A_i x + b_i)}{a_{0i}^T x + b_{0i}}, & j = m_i + 1, \end{cases}$$

so $D_{jj} \geq 0$ for $j = 1, \ldots m_i + 1$ matches the constraints in (3). The semidefinite constraint can be relaxed by adding a positive-definite diagonal matrix to the left-hand side:

$$\begin{bmatrix} (a_{0i}^T x + b_{0i})I & A_i x + b_i \\ (A_i x + b_i)^T & a_{0i}^T x + b_{0i} \end{bmatrix} + \xi_i \hat{I} \succeq 0, \qquad \xi_i \geq 0,$$

where $\hat{I}$ is the $(m_i + 1) \times (m_i + 1)$ identity matrix. The first two inequalities in (4) correspond to nonnegativity requirements on the entries of the diagonal matrix in the $LDL^T$ factorization of this matrix. The third inequality in (4) exactly matches the nonnegativity of $\xi$ to ensure that this is indeed a relaxation.

One advantage of this relaxation formulation over the one presented in [13] and [14] is that we only use $m$ relaxation variables instead of $2m$. Doing so means that we will have not only fewer variables but also fewer penalty parameters to control in the resulting primal–dual penalty problem.

Thus, the *primal penalty problem* can be formulated as

$$\begin{aligned} \min_{x,\xi} \quad & c^T x + d^T \xi \\ \text{s.t.} \quad & \frac{(A_i x + b_i)^T (A_i x + b_i)}{a_{0i}^T x + b_{0i} + \xi_i} \leq a_{0i}^T x + b_{0i} + \xi_i, & i = 1, \ldots, m, \\ & a_{0i}^T x + b_{0i} + \xi_i \geq 0, & i = 1, \ldots, m, \\ & a_{0i}^T x + b_{0i} \leq u_i, & i = 1, \ldots, m, \\ & \xi_i \geq 0, & i = 1, \ldots, m, \end{aligned} \tag{5}$$

where $d$ and $u$ are the strictly positive primal and dual penalty parameters, respectively. As discussed in [13, 14], relaxing a constraint in the primal problem leads to the primal penalty parameter of the relaxation acting as an upper bound on the dual variables. In order to establish a similar relaxation on the dual side, we introduce an upper bound on the primal side, and, again, this upper bound ends up serving as the dual penalty parameter of the dual relaxation. In fact, the dual problem has the following form:

$$\max_{y_0, y, \psi} -\sum_{i=1}^{m}(b_i^T y_i + b_{0i} y_{0i} + u_i \psi_i)$$

$$\text{s.t.} \quad \sum_{i=1}^{m}(A_i^T y_i + a_{0i} y_{0i}) \qquad\qquad = c,$$

$$y_0 + \psi \qquad\qquad\qquad \le d, \tag{6}$$

$$y_0 + \psi \qquad\qquad\qquad \ge 0,$$

$$\frac{y_i^T y_i}{y_{0i} + \psi_i} \qquad\qquad \le y_{0i} + \psi_i, i = 1, \ldots, m,$$

$$\psi \qquad\qquad\qquad \ge 0,$$

where $y_i \in \mathcal{R}^{m_i}, i = 1, \ldots, m$ and $y_0 \in \mathcal{R}^m$ are the dual variables and $\psi \in \mathcal{R}^m$ are the dual relaxation variables.

Note that for sufficiently large $d$ and $u$, both (5) and (6) have strictly feasible interiors. For the primal problem, we can pick any $x$, set $u$ to satisfy $a_{0i}^T x + b_{0i} < u_i$ for $i = 1, \ldots, m$, and we can let

$$\xi_i > \max\{0, -(a_{0i}^T x + b_{0i}), \|A_i x + b_i\| - (a_{0i}^T x + b_{0i})\}.$$

Similarly, for the dual problem, pick any $y$ and set $y_0$ in order to satisfy the first constraint of (6). (Since we no longer require $y_0 \ge 0$, it is possible to do so.) Then, we can pick any

$$\psi_i > \max\{0, -y_{0i}, \|y_i\| - y_{0i}\}$$

and set $d_i > y_{0i} + \psi_i$ for $i = 1, \ldots, m$.

Having strictly feasible interiors for both the primal and the dual problems means that both (5) and (6) have optimal solutions, and there is no duality gap. Thus, the pair (5) and (6) satisfy the regularity assumptions of standard interior-point algorithms for both SOCP and general NLP [1, 12].

Nevertheless, even though (5) and (6) exhibit regularity, the original SOCP may not. In fact, as it quite often happens within a branch-and-bound framework, the original SOCP may not even be feasible. It is shown in [12] that a solution with a duality gap, if it exists, can be recovered as the penalty parameters (either the primal or the dual, while keeping the other fixed) tend to infinity. Similarly, it is well known that the original objective function can be dropped and a *feasibility problem* can be solved as needed. One advantage of having a relaxation/penalty scheme for both the primal and the dual problems is that a feasibility problem can be designed for either one, in order to detect primal or dual infeasibility for the original SOCP.

### 3.3  A Primal–Dual Penalty Interior-Point Method

Since we will solve the pair (5) and (6) as NLPs, we will now describe the application of a standard interior-point method to these problems. This method, along with approaches to manage the penalty parameters, has been discussed extensively in [14] for a general NLP, so we will only provide a brief outline here, adapted to the case of a reformulated SOCP. Since the relaxed constraint (4) looks slightly different than the relaxed constraint in [14], we will need to introduce the appropriate first-order conditions, but the general outline of the overall solution method will be the same.

We start by introducing some auxiliary variables that will help simplify our formulation:

$$
\begin{aligned}
\min_{x,\xi,f,g} \quad & c^T x + d^T \xi \\
\text{s.t.} \quad f_i &= A_i x + b_i, & i &= 1,\ldots,m, \\
g_i &= a_{0i}^T x + b_{0i} + \xi_i, & i &= 1,\ldots,m, \\
g_i - \frac{f_i^T f_i}{g_i} &\geq 0, & i &= 1,\ldots,m, \\
g_i &\geq 0, & i &= 1,\ldots,m, \\
u_i - g_i + \xi_i &\geq 0, & i &= 1,\ldots,m, \\
\xi_i &\geq 0, & i &= 1,\ldots,m,
\end{aligned}
\tag{7}
$$

where $f_i \in \mathcal{R}^{m_i}$ and $g_i \in \mathcal{R}, i = 1,\ldots,m$ are the auxiliary variables. Since the first two constraints that serve to introduce these variables are affine equality constraints, (7) remains a convex nonlinear programming problem.

Formulating the log-barrier problem for (7),

$$
\begin{aligned}
\min_{x,\xi,f,g} \quad & c^T x + d^T \xi - \mu \sum_{i=1}^{m} \left( \log\left( g_i - \frac{f_i^T f_i}{g_i} \right) + \log g_i + \log(u_i - g_i + \xi_i) + \log \xi_i \right) \\
\text{s.t.} \quad & f_i = A_i x + b_i, & i &= 1,\ldots,m, \\
& g_i = a_{0i}^T x + b_{0i} + \xi_i, & i &= 1,\ldots,m,
\end{aligned}
$$
(8)

where $\mu > 0$ is the barrier parameter.

The first-order conditions for this problem are

$$
\begin{aligned}
A_i x - f_i + b_i &= 0, & i &= 1,\ldots,m, \\
a_{0i}^T x + \xi_i - g_i + b_{0i} &= 0, & i &= 1,\ldots,m, \\
c - \sum_{i=1}^{m} A_i^T y_i - \sum_{i=1}^{m} a_{0i} y_{0i} &= 0 \\
\psi_i(u_i - g_i + \xi_i) &= \mu, & i &= 1,\ldots,m, \\
\xi_i(d_i - y_{0i} - \psi_i) &= \mu, & i &= 1,\ldots,m, \\
(y_{0i} + \psi_i)\left( g_i - \frac{f_i^T f_i}{g_i} \right) &= \mu, & i &= 1,\ldots,m, \\
\frac{y_{0i} + \psi_i}{g_i} f_i + y_i &= 0, & i &= 1,\ldots,m.
\end{aligned}
\tag{9}
$$

Note that the last condition implies the second-order cone constraint in (6) since we would have that

$$y_i^T y_i = (y_{0i} + \psi_i)^2 \frac{f_i^T f_i}{g_i^2}$$

and $\frac{f_i^T f_i}{g_i^2} \leq 1$ in each iteration.

The first-order conditions are solved using Newton's method while performing a linesearch to guarantee progress toward optimality and modifying the value of $\mu$ at each iteration (see [14] or [15] for details). Of course, we need to also control the penalty parameters to guarantee that we have found a solution for the original SOCP or provide a certificate of infeasibility. In [14], Benson and Shanno discuss two approaches, static and dynamic updating, to resolve this issue.

- For static updating, the values of $d$ and $u$ are kept constant, and the problem is solved to optimality. Then, if $\xi > 0$ (or $\psi > 0$) at the optimal solution, the primal (or the dual) penalty parameters are increased and the new problem is solved. After a fixed number of updates are performed, the problem is declared a candidate for infeasibility. If another update is necessary, $c$ is set to 0 before solving the system again to detect primal infeasibility (or $b$ is set to 0 to detect dual infeasibility). If a feasible solution (for the original SOCP) is obtained at the end of this process, we return to solving (9) with higher values of the penalty parameters. Otherwise, we declare the problem to be infeasible.
- For dynamic updating, the progress of $g_i + \xi_i$ and $y_{0i} + \psi_i$ for $i = 1, \ldots, m$ toward their upper bounds of $u_i$ and $d_i$, respectively, is monitored at each iteration. If any of them are too close to their upper bounds, those bounds are increased. If any single bound is increased more than a fixed number of times, we modify the corresponding problem as described in static updating to enter the infeasibility detection phase. Similarly, if a feasible solution is found, we return to solving the original problem. Otherwise, we declare the problem to be infeasible.

While the static update is rather straightforward, it may require the complete solution of multiple problems. Therefore, as was the case in [14], the dynamic updating approach is preferred here as well.

In addition to its warmstarting capabilities, the primal–dual penalty approach also allows us to (approximately) solve SOCPs that have duality gaps at the optimal pair of primal–dual solutions. This asymptotic behavior of the relaxed problem is analyzed in [12].

## 3.4  Warmstarting

Most successful implementations for mixed-integer linear programming either use a simplex-type method to solve the underlying linear programming problems, or they use a crossover approach which starts simplex iterations and crosses over to an interior-point method as needed. This is due to the fact that a simplex-type method (or an active-set approach in nonlinear programming) is quite easy to restart from a previous solution. In contrast, starting an interior-point method from the optimal solution of another problem causes issues due to at least one of a complementary pair of primal–dual variables already being at its bound. A thorough analysis of the numerical difficulties is presented in [13, 14] for general linear and nonlinear programming warmstarts, respectively, and in [10,11] for warmstarts within branch-and-bound and outer approximation frameworks, respectively, for mixed-integer nonlinear programming. In all instances, it is shown that a standard interior-point method, applied directly to the original problem, will not only fail to warmstart but fatally stall if initialized from the optimal solution of a previously solved problem.

As pointed out in these papers, the primal–dual penalty approach serves as a remedy to the stalling issue by un-stalling the iterates and even improves on the iteration count over a coldstart. This is attained by keeping the optimal values for the primal–dual variables $x$, $g$, $y$, and $y_0$, but slightly perturbing the primal–dual relaxation variables $\xi$ and $\psi$ away from 0 (and recomputing $f$). This perturbation can be quite small ($10^{-4}$ usually suffices), since both $\xi$ and $\psi$ are variables and their values can increase as needed. This framework avoids stalling by moving all the terms of the complementarity conditions in (9) away from 0, but still close to the central path for a small value of $\mu$.

## 3.5  *Handling the Discrete Variables*

For our numerical experiments, we have implemented both a branch-and-bound method [40] and an outer approximation method [29] for a generic MINLP. Branch-and-bound conducts a search through a tree where each node is obtained by adding a bound to its parent to eliminate a noninteger solution and where each node requires the solution of a continuous NLP. Outer approximation alternates between the solution of an NLP obtained by fixing the integer variables and of an MILP obtained using linearizations of the objective function and the constraints at the solutions of the NLP. These methods and their use in conjunction with the primal–dual penalty interior-point method were analyzed in [10, 11]. We refer the reader to these papers for further details.

## 4  MISOCPs Arising in Portfolio Optimization

In this paper, we consider a single-period portfolio optimization problem which is based on the Markowitz mean–variance framework [47], where there is a trade-off between expected return and risk (market volatility) that the investor may be willing to take on. Portfolio optimization literature has come quite far in the decades since the publication of [46], and many modern models are formulated using second-order cone constraints and take discrete decisions into consideration. In this paper, we present a portfolio optimization model that is formulated as an MISOCP and is aggregated from the following in literature:

- About two decades ago, transaction costs were started to be taken into account by [24, 31, 48] in portfolio optimization problems. In [62], Yoshimoto models V-shaped transaction cost function in the mean–variance portfolio optimization framework. Transaction costs also have been studied by [25, 28, 37, 38] in this framework. We will use quadratic cost functions for the single-period model as proposed in [30] where the authors derive an optimal closed-form dynamic portfolio policy with predictable returns and transaction costs.

- In [42], Lobo et al. consider a single-period portfolio selection problem with linear and fixed transaction costs. They introduce a shortfall risk constraint in order to ensure that the terminal wealth is greater than a predetermined threshold level. They obtain second-order cone constraint from this formulation. They allow short selling in their model. By using linear transaction costs, they obtain a convex optimization problem which can be solved by using the general purpose software SOCP. When they introduce fixed transaction costs into their framework, their model is no longer convex. They relax their transaction cost constraint in order to obtain a convex problem and solve the relaxed problem by using branch-and-bound method. Their second approach is to provide an iterative heuristic. They obtain a suboptimal solution with this method by solving a small number of convex optimization problems. They show that there is a small gap between the suboptimal heuristic solution and the guaranteed upper bound with computational experiments. They suggest that these two methods can be incorporated for further accuracy levels.

- In [18], Bonami and Lejeune study a single-period portfolio optimization problem under stochastic and integer constraints as an extension of the classical mean–variance portfolio optimization framework. First they introduce a probabilistic portfolio optimization model where expected asset returns are stochastic and then they obtain their deterministic equivalents of these models to test different probability distributions to determine which ones provide an exact or approximate closed-form solution. They focus on different types of constraints that traders should take into account when they construct their portfolio, such as diversification by sectors, buy-in threshold, and round-lot constraints. These constraints were studied by [36] in absence of uncertainty about a decade ago. These sets of constraints provide binary and integer variables. They use branch-and-bound with two new proposed branching rules: *idiosyncratic risk* and *portfolio risk* branching. Numerical results are presented up to 200 assets and compare standard MINLP solvers and [17]. They suggest that the portfolio risk branching rule performs best in terms of robustness and speed.

In our model, we have included transaction costs, conditional value-at-risk (CVaR) constraints, and diversification-by-sector constraints. These model components/features have been adapted from [18, 30, 42].

The portfolio optimization model considered in this paper can be written as

$$
\max_{x^+, x^-, \zeta} \sum_{j=0}^{n} r_j (w_j + x_j^+ - x_j^-) - \frac{1}{2}(x^+ + x^-)^T \Lambda (x^+ + x^-)
$$

s.t.   $\Phi^{-1}(\eta_k) \|\Sigma^{\frac{1}{2}}(w + x^+ - x^-)\| \leq \sum_{j=0}^{n} r_j (w_j + x_j^+ - x_j^-) - W_k^{\mathrm{low}}, \quad k = 1, \ldots, M,$

$s_{\min}\zeta_k \leq \sum_{j \in S_k}(w_j + x_j^+ - x_j^-) \leq s_{\min} + (1 - s_{\min})\zeta_k, \quad k = 1, \ldots, L,$

$\sum_{k=1}^{L} \zeta_k \geq L_{\min},$

$\sum_{j=0}^{n}(w_j + x_j^+ - x_j^-) = 1,$

$w_j + x_j^+ - x_j^- \geq -s_j, \quad j = 1, \ldots, n,$

$x^+, x^- \geq 0,$

$\zeta \in \{0, 1\}^L,$

(10)

where we consider cash (index 0) and $n$ risky assets from $L$ different sectors for inclusion in our portfolio. The decision variables are $x^+$, $x^-$, and $\zeta$. Here, $x^+ \in \mathcal{R}^{n+1}$ and $x^- \in \mathcal{R}^{n+1}$ denote the buy and sell transactions, respectively. $\zeta \in \{0, 1\}^L$ is the vector whose elements denote whether there are sufficient investments in each sector. We describe the remaining model components in detail below.

## 4.1 Objective Function

The investor's objective is to choose the optimal trading strategies to maximize the end-of-period expected total return. Denoting the expected rates of return by $r \in \mathcal{R}^{n+1}$ and the current portfolio holdings by $w \in \mathcal{R}^{n+1}$, the expected total portfolio value at the end of the period is given by

$$
\sum_{j=0}^{n} r_j (w_j + x_j^+ - x_j^-).
$$

However, both the buy and sell transactions are penalized by transaction costs. According to recent dynamic portfolio choice literature (e.g., [20, 30]), transaction costs include a number of factors, such as price impacts of transactions, brokerage commissions, bid–ask spreads, and taxes. As such, there are a number of different ways to model transaction costs, including linear and convex or concave nonlinear cost functions. In this paper, we have decided to use the quadratic convex transaction cost formulation of [30] as it provides best fit to our framework. Therefore, the total transaction costs appear as a penalty term in the objective function:

$$
\frac{1}{2}(x^+ + x^-)^T \Lambda (x^+ + x^-),
$$

where $\Lambda \in \mathcal{R}^{(n+1)\times(n+1)}$ is the trading cost matrix and is obtained as a positive multiple of the covariance matrix of the expected returns. Because of this connection to a covariance matrix, $\Lambda$ is symmetric and positive definite. Note that both buy and sell transactions receive the same transaction cost, but it would be straightforward to instead include two quadratic terms in the objective function with different trading cost matrices for each type of transaction.

As we will see in the following discussion, the continuous relaxation of (10) includes only linear and second-order cone constraints. However, the quadratic term in the objective function prevents the overall problem from being formulated as an MISOCP. While we could simply classify the problem as a MINLP, we choose to instead reformulate it as an MISOCP so that the relaxation scheme described in the previous section can be applied to the existing second-order cone constraint. We introduce a new variable $\rho \in \mathcal{R}$ and rewrite the objective function of (10) as

$$\sum_{j=0}^{n} r_j (w_j + x_j^+ - x_j^-) - \rho,$$

with

$$\frac{1}{2}(x^+ + x^-)^T \Lambda (x^+ + x^-) \leq 2\rho.$$

Note that this constraint is equivalent to

$$(x^+ + x^-)^T \Lambda (x^+ + x^-) \leq (1 + \rho)^2 - (1 - \rho)^2,$$

and moving the last term to the left-hand side and taking the square root of both sides give the following second-order cone constraint:

$$\left\| \begin{pmatrix} \Lambda^{\frac{1}{2}}(x^+ + x^-) \\ 1 - \rho \end{pmatrix} \right\| \leq 1 + \rho.$$

$\Lambda^{\frac{1}{2}}$ exists since $\Lambda$ is positive definite. Additionally, this conversion does not increase the difficulty of solving the problem significantly—we add only one auxiliary variable, so the Newton system does not become significantly larger. Also, worsening the sparsity of the problem is not a concern here, since the original problem (10) has a quite dense matrix in the Newton system due to the covariance matrix and the related trading cost matrix both being dense.

## 4.2   Shortfall Risk Constraint

As stated above, we are considering both return and risk in this model. In the objective function, we focus on maximizing the expected total return less transaction costs, so we will seek to limit our risk using constraints. To that end, we will use CVaR constraints, as was done by Lobo et al. in [42].

For each CVaR constraint $k$, $k = 1, \ldots, M$, we will require that our expected wealth at the end of the period be above some threshold level $W_k^{\text{low}}$ with a probability of at least $\eta_k$. Thus, letting

$$W = \sum_{j=0}^{n} \hat{r}_j (w_j + x_j^+ - x_j^-),$$

where $\hat{r}$ is the random vector of returns, we require that

$$\mathcal{P}(W \geq W_k^{\text{low}}) \geq \eta_k, \qquad k = 1, \ldots, M.$$

We assume that the elements of $r$ have jointly Gaussian distribution so that W is normally distributed with a mean of

$$\sum_{j=0}^{n} r_j (w_j + x_j^+ - x_j^-)$$

and a standard deviation of

$$\|\Sigma^{\frac{1}{2}} (w + x^+ - x^-)\|,$$

where $\Sigma$ is the covariance matrix of the returns. As shown in [18], we can expand this assumption to a more general class of probability distributions, including symmetric probability distributions and positively skewed probability distributions. In fact, [18] also shows that our proposed model can approximate an even greater class of distributions, which encompasses any distribution that can be characterized by its first two moments.

Therefore, the CVaR constraints can be formulated as

$$\mathcal{P} \left( \frac{W - \sum_{j=0}^{n} r_j (w_j + x_j^+ - x_j^-)}{\|\Sigma^{\frac{1}{2}} (w + x^+ - x^-)\|} \geq \frac{W_k^{\text{low}} - \sum_{j=0}^{n} r_j (w_j + x_j^+ - x_j^-)}{\|\Sigma^{\frac{1}{2}} (w + x^+ - x^-)\|} \right) \geq \eta_k,$$

for each $k = 1, \ldots, M$. This implies that

$$1 - \Phi \left( \frac{W_k^{\text{low}} - \sum_{j=0}^{n} r_j (w_j + x_j^+ - x_j^-)}{\|\Sigma^{\frac{1}{2}} (w + x^+ - x^-)\|} \right) \geq \eta_k, \qquad k = 1, \ldots, M,$$

where $\Phi$ is the cumulative distribution function for a standard normal random variable, or

$$\Phi(z) = \frac{1}{\sqrt{2\pi}} \int_{-\infty}^{z} e^{-t^2/2} dt.$$

Rearranging the terms and taking the inverse give us

$$\frac{W_k^{\text{low}} - \sum_{j=0}^{n} r_j (w_j + x_j^+ - x_j^-)}{\|\Sigma^{\frac{1}{2}} (w + x^+ - x^-)\|} \leq \Phi^{-1} (1 - \eta_k), \qquad k = 1, \ldots, M.$$

Using the symmetry of the standard normal distribution function, we can rewrite the constraint again as

$$-\frac{W_k^{\text{low}} - \sum_{j=0}^{n} r_j(w_j + x_j^+ - x_j^-)}{\|\Sigma^{\frac{1}{2}}(w + x^+ - x^-)\|} \geq \Phi^{-1}(\eta_k), \qquad k = 1, \ldots, M.$$

Finally, rearranging the terms gives us the second-order cone constraint in (10)

$$\Phi^{-1}(\eta_k)\|\Sigma^{\frac{1}{2}}(w + x^+ - x^-)\| \leq \sum_{j=0}^{n} r_j(w_j + x_j^+ - x_j^-) - W_k^{\text{low}}, \qquad k = 1, \ldots, M.$$

## 4.3 Diversification by Sectors

Diversification is another important instrument used to reduce the level of risk in the portfolio. In this part, we impose a diversification requirement to the investor to allocate sufficiently large amounts in at least $L_{\min}$ of the $L$ different economic sectors. This type of constraint was considered by [18].

To express this diversification requirement, we start by defining binary variables $\zeta_k \in \{0, 1\}, k = 1, \ldots, L$ for each economic sector $k$. If $\zeta_k = 1$, our total portfolio allocation in assets from sector $k$ will be at least $s_{\min}$ (and, of course, no more than 1). Otherwise, it will mean that our total portfolio allocation in those assets fell short of the threshold level $s_{\min}$. We can express these requirements with a constraint in the following form:

$$s_{\min}\zeta_k \leq \sum_{j \in S_k}(w_j + x_j^+ - x_j^-) \leq s_{\min} + (1 - s_{\min})\zeta_k, \quad k = 1, \ldots, L,$$

where $S_k$ is the set of assets that belong to economic sector $k, k = 1, \ldots, L$.

In order to express the diversification requirement, we also need to introduce a cardinality constraint:

$$\sum_{k=1}^{L} \zeta_k \geq L_{\min}. \tag{11}$$

## 4.4 Portfolio Constraints

The remaining constraints in our problem are grouped into the general category of portfolio constraints. The first of these,

$$\sum_{j=0}^{n}(w_j + x_j^+ - x_j^-) = 1,$$

requires that we allocate 100% of our portfolio at the end of the investment period. Since we start with

$$\sum_{j=0}^{n} w_j = 1,$$

this constraint can also be written as

$$\sum_{j=0}^{n} x_j^+ = \sum_{j=0}^{n} x_j^-,$$

which provides a balance between the buy and sell transactions.

Additionally, we have another constraint that allows for short sales of the nonliquid assets by stating that we can take a limited short position for each one:

$$w_j + x_j^+ - x_j^- \geq -s_j, \qquad j = 1, \ldots, n,$$

where $s$ represents the short position limit for each nonliquid asset.

Finally, we require that $x^+$ and $x^-$, the variables associated with the buy and sell transactions must be nonnegative.

With the modifications to the model due to the transaction costs, the MISOCP we will be solving in our numerical testing will have the form

$$\max_{x^+, x^-, \zeta, \rho} \sum_{j=0}^{n} r_j (w_j + x_j^+ - x_j^-) - \rho$$

s.t.
$$\left\| \begin{pmatrix} \Lambda^{\frac{1}{2}}(x^+ + x^-) \\ 1 - \rho \end{pmatrix} \right\| \leq 1 + \rho,$$

$$\Phi^{-1}(\eta_k) \| \Sigma^{\frac{1}{2}} (w + x^+ - x^-) \| \leq \sum_{j=0}^{n} r_j (w_j + x_j^+ - x_j^-) - W_k^{\text{low}}, \quad k = 1, \ldots, M,$$

$$s_{\min} \zeta_k \leq \sum_{j \in S_k} (w_j + x_j^+ - x_j^-) \leq s_{\min} + (1 - s_{\min}) \zeta_k, \quad k = 1, \ldots, L,$$

$$\sum_{k=1}^{L} \zeta_k \geq L_{\min},$$

$$\sum_{j=0}^{n} (w_j + x_j^+ - x_j^-) = 1,$$

$$w_j + x_j^+ - x_j^- \geq -s_j, \qquad j = 1, \ldots, n,$$

$$x^+, x^- \geq 0$$

$$\zeta \in \{0, 1\}^L.$$

$$(12)$$

As mentioned above, there are two additional types of constraints appearing in literature that we would like to include in future testing. The first of these, buy-in threshold constraints, requires additional binary variables and linear constraints and therefore keeps the problem as an MISOCP. We did not include them in this paper since we already have an MISOCP and the particular data set we chose led to either infeasible or trivially solved problems when the buy-in thresholds were added. The second type of constraint, round-lot constraints, could introduce nonlinear functions into our constraints, and we wanted to focus on MISOCPs in this paper and leave MINLPs with second-order cone constraints for future work.

### 4.5 Smoothness of the Second-Order Cone Constraints

As discussed in Sect. 3, we picked the ratio reformulation in order to guarantee that the underlying SOCPs would be smooth. We will now examine this choice for the second-order cone constraints included in (12).

For the transaction cost constraints, the right-hand side term is $1 + \rho$. Since the total transaction cost paid will be $2\rho$, we have that $\rho \geq 0$. Therefore, $1 + \rho \geq 1$, and the right-hand side is bounded away from 0.

For the shortfall constraints, note that we start with $\sum_{j=0}^n w_j = 1$ and that, since we are focusing on shortfalls, $W_k^{\text{low}} < 1$. Also note that if we assume that our initial asset allocation satisfies the diversification by sector constraints, we can define a feasible solution that does not require us to buy or sell any assets. Our objective is to maximize our end-of-period expected total return, which means that we expect our optimal allocation to do at least as well as this feasible solution. Thus, we can guarantee that

$$\sum_{j=0}^n r_j (w_j + x_j^+ - x_j^-) \geq 1 > W_k^{\text{low}}, \qquad k = 1, \dots, M,$$

which means that the right-hand side is bounded away from 0.

With these reformulations, the first two constraints in (12) can be rewritten as

$$\frac{(x^+ + x^-)^T \Lambda (x^+ + x^-) + (1 - \rho)^2}{1 + \rho} \leq 1 + \rho,$$

$$\left\| \frac{(\Phi^{-1}(\eta_k))^2 (w + x^+ - x^-)^T \Sigma (w + x^+ - x^-)}{\sum_{j=0}^n r_j (w_j + x_j^+ - x_j^-) - W_k^{\text{low}}} \right\| \leq \sum_{j=0}^n r_j (w_j + x_j^+ - x_j^-) - W_k^{\text{low}}, \quad k = 1, \dots, M.$$

## 5 Numerical Results

In our numerical testing, we consider one riskless and 20–400 risky assets for trading. The risky assets are chosen from the S&P500 list of companies in alphabetical order, and each stock is matched with its real-world economic sector. The geometric mean and the covariance of the risky assets were calculated from the closing prices of the stocks in 2010. The riskless asset which refers to investment in the money market has a 1% return.

**Table 1** Results of the branch-and-bound algorithm

| | | | Coldstart | | Warmstart | | |
|---|---|---|---|---|---|---|---|
| $n$ | $L$ | $M$ | Nodes | Iters | Nodes | Iters | % Impr |
| 20 | 6 | 2 | 7 | 111 | 7 | 63 | 43.2 |
| 50 | 10 | 2 | 25 | 424 | 27 | 282 | 33.5 |
| 100 | 10 | 2 | 33 | 705 | 33 | 446 | 36.7 |
| 200 | 10 | 2 | 11 | 261 | 11 | 184 | 29.5 |
| 400 | 10 | 2 | 19 | 527 | 11 | 238 | 22.0 |

As we discussed before, we follow both [18, 42] formulation in our framework. Therefore, we generally use the same constraint parameters with these two studies for consistency.

**Initial weights for the stocks:** $w_j = 1/(n+1), j = 0, \ldots, n$
**Shortfall risk constraint parameters:** $\eta_1 = 95\%, W_1^{\text{low}} = 0.90, \eta_2 = 99.7\%,$
$W_2^{\text{low}} = 0.95$
**Diversification by sectors parameters:** $L_{\min} = \lceil 0.5 \times L \rceil, s_{\min} = 0.01$
**Short sale portfolio constraints:** $s_j = 0.5/n, j = 1, \ldots, n, s_0 = 0.5$

The problem instances are modeled using Matlab Version 2009b and solved on a machine with 2.53GHz Intel Core 2 Duo Processors, 4GB of main memory, and running Mac OS X Version 10.6.8. We solve our problem instances using the Matlab-based solver MILANO [9] Version 1.4 which implements both branch-and-bound and outer approximation algorithms and uses the primal–dual penalty interior-point approach that allows warmstarting, as described in Sect. 3. Table 1 illustrates the result of the branch-and-bound algorithm while Table 2 presents the results of the outer approximation algorithm. The first column ($n$) is the number of assets considered for the instance, the second column ($L$) is number of different economic sectors, and the third column ($M$) gives the number of CVaR constraints included in the model. The next four columns show the numbers of nodes and iterations that are required to solve the problem after either a coldstart or a warmstart. The last column represents the percentage improvement in the average number of iterations per node, as attained by warmstarting, and the numbers show that we obtain substantial improvements by using warmstarting for both the branch-and-bound and outer approximation algorithms.

There are a few instances where the total number of nodes taken by the algorithm after a coldstart differs from that after a warmstart. This can be due to the existence of multiple optima or simply to differences in accuracy of the solutions obtained by the two approaches that may lead to a different branching pattern. That is why we have chosen to use the average number of iterations per node, instead of the total, in our improvement calculations. For example, while the warmstart code takes a total of 238 iterations on the $n = 400$ instance in Table 1, as compared to the 527 iterations of the coldstart code, it does so after solving only 11 NLP subproblems, instead of 19. Therefore, the warmstart uses an average of 21.64 iterations per

**Table 2** Results of the outer approximation algorithm

|   |   |   | Coldstart | | Warmstart | | |
|---|---|---|---|---|---|---|---|
| $n$ | $L$ | $M$ | Nodes | Iters | Nodes | Iters | % Impr |
| 20 | 6 | 2 | 2 | 36 | 2 | 31 | 13.9 |
| 50 | 10 | 2 | 3 | 65 | 3 | 52 | 20.0 |
| 100 | 10 | 2 | 3 | 89 | 3 | 60 | 32.3 |
| 200 | 10 | 2 | 2 | 57 | 3 | 69 | 27.7 |
| 400 | 10 | 2 | 2 | 69 | 2 | 53 | 23.2 |

subproblem, whereas the coldstart uses 27.74. This is an improvement of roughly 22% per node. We believe that this is a fair way to assess the impact of the proposed warmstart approach.

In Table 2, we can see that the number of nodes needed to solve the portfolio optimization problem is either 2 or 3. We have tried to create a new test set that would exhibit additional iterations of the outer approximation approach, but were not successful in doing so. This phenomenon seems to be related to the structure of the current problem, with the only nonlinearities being 3 second-order cone constraints, and not related to this particular instance of the problem. Preliminary testing on the multiperiod version of this model using the same data has indeed exhibited more iterations.

It should be noted here that the solver MILANO is relatively new and lacks sophistication in its linear algebra routines at this stage of its development. It is our sincere hope that in future work, we will be able to provide favorable comparisons to existing software for MISOCP, including CPLEX.

# 6 Conclusion

In this paper, we presented a set of techniques for solving MISOCPs as MINLPs whose underlying NLPs are smooth, regularized, and convex. A ratio reformulation was used to smooth the underlying SOCPs. The primal–dual penalty interior-point method, modified from that presented in [10, 11], was then used to provide warmstarts, regularization, and infeasibility detection capabilities, and the modification also exploited the structure of the MISOCP. We have implemented both branch-and-bound and outer approximation frameworks that use this method and use them to solve portfolio optimization problems. Numerical results show that we can solve small- to medium-sized instances successfully. The infeasibility detection capability provided by the primal–dual penalty approach allows us to either solve or declare infeasibility at each node, thereby leading to a robust method. The warmstart capability is shown to significantly improve algorithm efficiency.

In future work, we hope to extend our approach to general MISOCPs by having a dynamic choice of constraint reformulations to resolve nonsmoothness issues. For handling the integer variables, our proposed frameworks can accommodate the various cuts appearing in MISOCP literature, and we will investigate such algorithmic improvements as well. Additionally, we will continue our work on portfolio optimization models by working to include buy-in thresholds and round-lot constraints in our models and extending to a multiperiod model.

# References

1. Alizadeh, F., Goldfarb, D.: Second-order cone programming. Math. Program. **95**, 3–51 (2003)
2. Andersen, E., Andersen, K.: The MOSEK optimization software. EKA Consulting ApS, Denmark (2000)
3. Atamturk, A., Berenguer, G., Max Shen, Z.-J.: A conic integer programming approach to stochastic joint location-inventory problems. Oper. Res. **60**, 366–381 (2012)
4. Atamturk, A., Narayanan, V.: Cuts for conic mixed-integer programming. In: Fischetti, M., Williamson, D.P. (eds.) Proceedings of IPCO 2007, pp. 16–29. Springer-Verlag Berlin Heidelberg (2007)
5. Atamturk, A., Narayanan, V.: Conic mixed-integer rounding cuts. Math Program. Ser. A **122**, 1–20 (2010)
6. Balas, E., Ceria, S., Cornuejols, G.: A lift-and-project cutting plane algorithm for mixed 0-1 programs. Math. Program. **58**, 295–324 (1993)
7. Belotti, P., Goez, J.C., Polik, I., Ralphs, T.K., Terlaky, T.: A conic representation of the convex hull of disjunctive sets and conic cuts for integer second order cone optimization. Technical report (2012)
8. Ben-Tal, A., Nemirovski, A.: On polyhedral approximations of the second-order cone. Math. Oper. Res. **26**(2), 193–205 (2001)
9. Benson, H.Y.: MILANO - a Matlab-based code for mixed-integer linear and nonlinear optimization. http://www.pages.drexel.edu/~hvb22/milano (2007)
10. Benson, H.Y.: Mixed-integer nonlinear programming using interior-point methods. Optim. Methods Softw. **26**(6), 911–931 (2011)
11. Benson, H.Y.: Using interior-point methods within an outer approximation framework for mixed integer nonlinear programming. In: Lee, J., Leyffer, S., (eds.) Mixed Integer Nonlinear Programming of IMA Volumes on Mathematics and Its Applications, vol. 154, pp. 225–243. Springer, New York (2012)
12. Benson, H.Y., Sen, A., Shanno, D.F.: Convergence analysis of an interior-point method for nonconvex nonlinear programming. Technical report (2009)
13. Benson, H.Y., Shanno, D.F.: An exact primal-dual penalty method approach to warmstarting interior-point methods for linear programming. Comput. Optim. Appl. **38**(3), 371–399 (2007)
14. Benson, H.Y., Shanno, D.F.: Interior-point methods for nonconvex nonlinear programming: Regularization and warmstarts. Comput. Optim. Appl. **40**(2), 143–189 (2008)
15. Benson, H.Y., Shanno, D.F., Vanderbeim, R.J.: A comparative study of large scale nonlinear optimization algorithms. In: Proceedings of the Workshop on High Performance Algorithms and Software for Nonlinear Optimization, Erice (2001)
16. Benson, H.Y., Vanderbei, R.J.: Solving problems with semidefinite and related constraints using interior-point methods for nonlinear programming. Math. Program. B **95**(2), 279–302 (2003)
17. Bonami, P., Biegler, L.T., Conn, A.R., Cornuejols, G., Grossman, I.E., Laird, C.D., Lee, J., Lodi, A., Margot, F., Sawaya, N., Waechter, A.: An algorithmic framework for convex mixed integer nonlinear programs. Discrete Optim. **5**, 186–204 (2008)
18. Bonami, P., Lejeune, M.A.: An exact solution approach for portfolio optimization problems under stochastic and integer constraints. Oper. Res. **57**(3), 650–670 (2009)

19. Brandenberg, R., Roth, L.: New algorithms for $k$-center and extensions. In: Yang, B., Du, D-Z., Wang, C.A. (eds.) Proceedings of COCOA 2008, pp. 64–78. Springer (2008)
20. Brown, D.B., Smith, J.E.: Dynamic portfolio optimization with transaction costs: Heuristics and dual bounds. Manag. Sci. 57(10), 1752–1770 (2011)
21. Cezik, M.T., Iyengar, G.: Cuts for mixed 0-1 conic programming. Math. Program. Ser. A 104, 179–202 (2005)
22. Cheng, Y., Drewes, S., Philipp, A., Pesavento, M.: Joint network optimization and beam-forming for coordinated multi-point transmission using mixed integer programming. In: Proceedings of ICASSP 2012, pp. 3217–3220. IEEE, New Jersey (2012)
23. Dadush, D., Dey, S.S., Vielma, J.P.: The split closure of a strictly convex body. Oper. Res. Lett. 39, 121–126 (2011)
24. Dantzig, G.B, Infanger, G.: Multi-stage stochastic linear programs for portfolio optimization. Ann. Oper. Res. 45(1), 59–76 (1993)
25. De Roon, F.A., Nijman, T.E, Werker, B.JM: Testing for mean-variance spanning with short sales constraints and transaction costs: the case of emerging markets. J. Financ. 56(2), 721–742 (2002)
26. Drewes, S.: Mixed integer second order cone programming. PhD Thesis. Technischen Universitat Darmstadt, Munich (2009)
27. Du, Y., Chen, Q., Quan, X., Long, L., Fung, R.Y.K.: Berth allocation considering fuel consumption and vessel emissions. Transport. Res. Part E 47, 1021–1037 (2011)
28. Dumas, B., Luciano, E.: An exact solution to a dynamic portfolio choice problem under transactions costs. J. Financ. 46(2), 577–595 (2012)
29. Duran, M.A., Grossmann, I.E.: An outer-approximation algorithm for a class of mixed-integer nonlinear programs. Math. Program. 36, 307–339 (1986)
30. Garleanu, N.B., Pedersen, L.H.: Dynamic trading with predictable returns and transaction costs. Technical report, National Bureau of Economic Research (2009)
31. Gennotte, G., Jung, A.: Investment strategies under transaction costs: the finite horizon case. Manag. Sci. 40(3), 385–404 (1994)
32. Geoffrion, A.M.: Generalized benders decomposition. J. Optim. Theory Appl. 10(4), 237–260 (1972)
33. Goldfarb, D.: The simplex method for conic programming. Technical Report TR-2002-05, CORC, IEOR Department of Columbia University, (2002)
34. Hijazi, H., Bonami, P., Ouorou, A.: Robust delay-constrained routing in telecommunications. Technical report, CNRS-Aix Marseille Universite (2012)
35. IBM/ILOG. CPLEX 11.0 reference manual. http://www.ilog.com/products/cplex/.
36. Jobst, N.J, Horniman, M.D, Lucas, C.A, Mitra, G.: Computational aspects of alternative portfolio selection models in the presence of discrete asset choice constraints. J. Quan. Financ. 1(5), 489–501 (2001)
37. Kellerer, H., Mansini, R., Speranza, M.G.: Selecting portfolios with fixed costs and minimum transaction lots. Ann. Oper. Res. 99(1), 287–304 (2000)
38. Konno, H., Wijayanayake, A.: Portfolio optimization problem under concave transaction costs and minimal transaction unit constraints. Math. Program. 89(2), 233–250 (2001)
39. Krishnan, K., Mitchell, J.: Linear programming approaches to semidefinite programming problems. Technical report, Department of Mathematical Sciences, Rensselaer Polytechnic Institute, Troy, NY, 12180 (2001)
40. Land, A.H., Doig, A.G.: An automatic method of solving discrete programming problems. Econometrica 28(3), 497–520 (1960)
41. Lasserre, J.B.: Global optimization with polynomials and the methods of moments. SIAM J. Optim. 1, 796–817 (2001)
42. Lobo, M.S., Fazel, M., Boyd, S.: Portfolio optimization with linear and fixed transaction costs. Ann. Oper. Res. 152(1), 341–365 (2007)
43. Lobo, M.S., Vandenberghe, L., Lebret, H., Boyd, S.: Applications of second-order cone programming. Linear Algebra Appl. 284(1–3), 13–228 (1998)

44. Lovasz, L., Schrijver, A.: Cones of matrices and set-functions and 0-1 optimization. SIAM J. Optim. **1**(2), 166–190 (1991)
45. Mak, H-Y., Rong, Y., Shen, Z-J.M.: Infrastructure planning for electric vehicles with battery swapping. Technical report, Submitted to Management Science (2012)
46. Markowitz, H.: Portfolio selection. J. Financ. **7**(1), 77–91 (1952)
47. Markowitz, H.M.: Portfolio Selection: Efficient Diversification of Investments. Wiley, New York (1959)
48. Mulvey, J.M., Vladimirou, H.: Stochastic network programming for financial planning problems. Manag. Sci. **38**(11), 1642–1664 (1992)
49. Nesterov, Y.E., Nemirovsky, A.S.: Interior Point Polynomial Methods in Convex Programming: Theory and Algorithms. SIAM Publications, Philadelphia (1993)
50. Ozsen, L., Coullard, C.R., Daskin, M.S.: Capacitated warehouse location model with risk pooling. Nav. Res. Logist. **55**, 295–312 (2008)
51. Pinar, M.C.: Mixed-integer second-order cone programming for lower hedging of American contingent claims in incomplete markets. Optim. Lett. **7**(1), 63–78 (2013) Online First
52. Shen, Z.J.M, Coullard, C.R., Daskin, M.S.: A joint location-inventory model. Transpor. Sci. **37**, 40–55 (2003)
53. Sherali, H.D., Adams, W.P.: A hierarchy of relaxations between the continuous and convex hull representations of zero-one programming problems. SIAM J. Discrete Math. **3**, 411–430 (1990)
54. Sherali, H.D., Adams, W.P.: A hierarchy of relaxations for mixed-integer zero-one programming problems. Discrete Appl. Math. **52**, 83–106 (1994)
55. Soberanis, P.A.: Risk optimization with $p$-order cone constraints. PhD Dissertation, University of Iowa (2009)
56. Stubbs, R.A., Mehrotra, S.: A branch-and-cut method for 0-1 mixed convex programming. Math. Program. Ser. A **86**, 515–532 (1999)
57. Sturm, J.F.: Using SeDuMi 1.02: a Matlab toolbox for optimization over symmetric cones. Optim. Methods Softw. **11–12**, 625–653 (1999). Version 1.05 available from http://fewcal.kub.nl/sturm
58. Taylor, J.A., Hover, F.S.: Convex models of distribution and system reconfiguration. IEEE Trans. Power Syst. **27**(3), 1407–1413 (2012)
59. Toh, K.C., Todd, M.J., Tutuncu, R.H.: SDPT3 — a matlab software package for semidefinite programming. Optim. Methods Softw. **11**, 545–581 (1999)
60. Vielma, J.P., Ahmed, S., Nemhauser, G.L.: A lifted linear programming branch-and-bound algorithm for mixed-integer conic quadratic programs. INFORMS J. Comput. **20**(3), 438–450 (2008)
61. Westerlund, T., Pettersson, F.: An extended cutting plane method for solving convex MINLP problems. Comput. Chem. Eng. **19**, 131–136 (1995)
62. Yoshimoto, A.: The mean-variance approach to portfolio optimization subject to transaction costs. J. Oper. Res. Soc. Jpn. **39**(1), 99–117 (1996)

# The Duality Between the Perceptron Algorithm and the von Neumann Algorithm

Dan Li and Tamás Terlaky

**Abstract** The perceptron and the von Neumann algorithms were developed to solve linear feasibility problems. In this paper, we investigate and reveal the duality relationship between these two algorithms. The specific forms of linear feasibility problems solved by the perceptron and the von Neumann algorithms are a pair of alternative systems by the Farkas Lemma. A solution of one problem serves as an infeasibility certificate of its alternative system. Further, we adapt an Approximate Farkas Lemma to interpret the meaning of an approximate solution from its alternative perspective. The Approximate Farkas Lemma also enables us to derive bounds for the distance to feasibility or infeasibility from approximate solutions of the alternative systems. Based on these observations, we interpret variants of the perceptron algorithm as variants of the von Neumann algorithm and vice versa, as well as transit the complexity results from one family to the other.

**Keywords** Linear feasibility problem • Perceptron algorithm • Von Neumann algorithm • Duality • Approximate Farkas Lemma

**Mathematics Subject Classification** 49N15, 90C05, 90C60, 68Q25

## 1 Introduction

Linear optimization (LO) is the problem of minimizing or maximizing a linear objective function subject to a system of linear inequalities and equations. The linear feasibility problem (LFP) is to find a feasible solution to a linear inequality system.

D. Li • T. Terlaky (✉)
Department of Industrial and Systems Engineering, Lehigh University, Bethlehem, PA, USA
e-mail: dal207@lehigh.edu; terlaky@lehigh.edu

L.F. Zuluaga and T. Terlaky (eds.), *Modeling and Optimization: Theory and Applications*, 113
Springer Proceedings in Mathematics & Statistics 62, DOI 10.1007/978-1-4614-8987-0_5,
© Springer Science+Business Media New York 2013

Considering an LO problem with zero as its objective function, then every feasible solution is optimal. From this point of view, the LFP is a special case of LO. On the other hand, it is well known [3, 24] that by the LO duality theorem, any LO problem can be transformed to an equivalent LFP. LFPs can be written in various forms. In this paper, we consider two of them. The first one is

$$A^T y > 0, \tag{1}$$

where matrix $A \in \mathbb{R}^{m \times n}$ with its column vectors $a_1, a_2, \ldots, a_n \in \mathbb{R}^m$ and $y \in \mathbb{R}^m$. We assume that $\|a_i\|_2 = 1$ for all $i = 1, 2, \ldots, n$. This assumption is not changing the status of feasibility of problem (1). The second form we consider is

$$Ax = 0, \quad e^T x = 1, \quad x \geq 0, \tag{2}$$

where $x \in \mathbb{R}^n$, $e \in \mathbb{R}^n$ is the vector of all one. Without loss of generality [1], we can assume that matrix $A$ has the same properties as in problem (1). Observe that problem (2) is a standard form of the LFP with a convexity constraint [9]. Let $\text{conv}(A)$ represent the convex hull of the points $a_i$. If the origin $0 \in \text{conv}(A)$, then problem (2) is feasible and can be considered as a weighted center problem [9], i.e., the problem of assigning nonnegative weights $x_i$ to the points $a_i$ so that their weighted center of gravity is the origin $0$.

Several algorithms are proposed for solving LFPs [3, 24, 26], such as simplex methods [3, 26], ellipsoid [18, 19] and interior point [24] methods, and variants of the perceptron [4, 5, 27] and the von Neumann [7, 9] algorithms. All of these algorithms aim to find a feasible solution to an LO, or equivalently to an LFP. They either deliver a feasible solution or provide an evidence of infeasibility. Ellipsoid and interior point methods are polynomial time algorithms while the perceptron and the von Neumann algorithms are not polynomial time [3, 18, 24, 26].

The perceptron algorithm [25] was originally invented in the field of machine learning. It is used to solve data classification problems by learning a linear threshold function. The von Neumann algorithm is published by Dantzig [7, 9] and it is a special case of the Frank–Wolfe algorithm [2, 13], which is an iterative method for solving constrained convex optimization problems. In this paper, we focus on the perceptron and the von Neumann algorithms. These two algorithms solve LFPs in the forms of (1) and (2), respectively. Therefore, we also call problem (1) the perceptron problem and problem (2) the von Neumann problem. By the Farkas Lemma, problems (1) and (2) are alternative systems to each other; and consequently, the perceptron and the von Neumann algorithms can be interpreted in a close duality relationship. This duality relationship allows us to transit one algorithm as a variant of the other, and we can transit their complexity results. For instance, the perceptron algorithm could be applied to an infeasible instance of problem (2), because its Farkas alternative system is a feasible LFP in the form of (1). The discovery of this duality not only provides a profound insight into both of the algorithms, but also results in new variants of the algorithms. In particular, variants of the von Neumann algorithm might lead to new versions of the Frank–Wolfe algorithm for solving convex optimization problems. From application point of view, the duality relationship of the two algorithms can be used to detect unexpected infeasibility caused by data error.

There is one major difference between these two algorithms. When problem (1) is feasible, the perceptron algorithm obtains an exact feasible solution—which satisfies all the constraints—with finite number of iterations. However, the von Neumann algorithm is an infinite algorithm. In any finite number of iterations, the solution returned is approximately feasible—the solution does not necessarily strictly satisfy the equality constraint $Ax = 0$. In order to identify an exact solution to problem (2), Dantzig [8] proposed a "bracketing" procedure: apply the von Neumann algorithm $m + 1$ times to finding approximate solutions to $m + 1$ perturbations of problem (2) and then combine these approximate solutions to get an exact solution to the original unperturbed problem. However in practice, this method cannot detect infeasibility, and it fails if problem (2) is close to the boundary of feasibility and infeasibility [8].

When we obtain an approximate solution without any knowledge of the feasibility status of the problem, we would like to explore the meaning of this approximate solution from the dual side and provide some quantified information about the feasibility of the dual problem. Todd and Ye [30] derived Approximate Farkas Lemmas from the gauge duality results of Freund [14] that are extensions of the Farkas Lemma [3, 24, 26]. We adapt an Approximate Farkas Lemma to problems (1) and (2). The Lemma serves as a basis of obtaining more meaningful information for the output of the algorithms. An $\epsilon$-solution of problem (2) implies that the distance of the corresponding problem (1) to infeasibility is no more than $\epsilon$.

This paper is structured as follows. In Sect. 2, we review the perceptron and von Neumann algorithms and their complexity results. Then we analyze the duality relationship between these two algorithms and discuss the use of Approximate Farkas Lemma in Sect. 3. In Sect. 4 and 5, we interpret variants of the perceptron algorithm as variants of the von Neumann algorithm and the other way around. The complexity results are also transferred from one family to the other. Finally, in Sect. 6 we state our conclusions and suggest avenues for future work.

## 2 Preliminaries

For convenience, we first introduce the following notations:

*The vector $x(y)$* – Let $\Delta_n$ be the unit simplex, i.e., $\Delta_n = \{x : x \geq 0, \|x\|_1 = 1\}$. For $y \in \mathbb{R}^m$, define $x(y) = \text{argmin} \{y^T Ax \mid x \in \Delta_n\}$. Thus, we have $Ax(y) = a_s$ or equivalently $x(y) = e_s$ if and only if $a_s^T y = \min \{a_i^T y \mid i = 1, \ldots, n\}$. Observe that $a_s$ and $x(y)$ are not necessarily unique.

*The condition number $\rho$* – A variety of condition numbers for linear optimization problems have been defined, e.g., by Renegar [22, 23], Epelman and Freund [12], Epelman and Vera [15, 16], and Cheung and Cucker [6]. In our analysis, we focus on one of them—$\rho$: a common feature of the perceptron and the von Neumann algorithms is that their iteration complexity depends on a condition number $\rho$ [4,11]

that quantifies how far the given instance is from the boundary of infeasibility and feasibility. Given an LFP, we define $\rho$ in general terms as follows.

**Definition 1.** Given an LFP, if the problem is feasible, $\rho$ is defined as the radius of the largest inscribed ball contained in the feasible region. Otherwise, $\rho$ measures the distance to feasibility.

If the problem is feasible (infeasible), the condition number $\rho$ indicates how far the constraints can be shifted or rotated so that the problem becomes infeasible (feasible). The larger $\rho$ is, the harder to turn a feasible (infeasible) problem to infeasible (feasible). Therefore, $\rho$ can be seen as a measurement of the robustness of feasibility or infeasibility. For different algorithms and problem forms, $\rho$ has slightly different definitions. We will discuss them in details in Sect. 3.2. In order to distinguish $\rho$ in the two different algorithms, we use $\rho_p(A)$ to represent $\rho$ in the perceptron algorithm and $\rho_v(A)$ for the von Neumann algorithm. In Sects. 2.1 and 2.2, we discuss that the condition number $\rho$ plays an essential role in the theoretical complexity of the perceptron and the von Neumann algorithms.

## 2.1   The Perceptron Algorithm

The Classical Perceptron Algorithm, used in machine learning [27], is designed to solve data classification problems: given a set of points with each point labeled as either positive or negative. The problem is to find a separating hyperplane, which separates all positive points from the negative ones. By some transformations [4,27] those problems lead to LFPs in the form of (1). The Classical Perceptron Algorithm assumes that problem (1) is strictly feasible; it starts from the origin and calls the classical perceptron update at each iteration. The classical perceptron update finds a violated constraint and moves the current iterate $y$ by one unit perpendicularly towards the violated constraint. The algorithm is described as follows.

**Algorithm 1:** The Classical Perceptron Algorithm

**Initialization** Let $y^0$ be the all-zero vector. $k = 0$.
**While True Do**
Find a column $a_j, j \in \{1, 2, \cdots, n\}$ such that $a_j^T y^k \leq 0$.
**If** such an $a_j$ does not exist, STOP and return $y^k$.
   **Else**

$$y^{k+1} = y^k + a_j,$$

$$k = k + 1.$$

**End While**

The perceptron algorithm has the following complexity result:

**Theorem 1 ([21]).** *Assume that the LFP (1) is strictly feasible. Then in at most*

$$\left\lceil \frac{1}{\rho_p(A)^2} \right\rceil$$

*iterations, the perceptron algorithm terminates with a feasible solution.*

## 2.2 The von Neumann Algorithm

The von Neumann algorithm, published by Dantzig [7, 9] in 1991, solves LFPs in the form of (2). Unlike the perceptron algorithm, the von Neumann algorithm gives an approximate solution in finite iterations. Thus, before presenting the von Neumann algorithm, we need to define what an $\epsilon$-*approximate solution* (or $\epsilon$-solution for short reference) of problem (2) is.

**Definition 2.** An $x \in \Delta_n$ is called an $\epsilon$-solution of problem (2) if $\|Ax\| \le \epsilon$.

The von Neumann algorithm terminates once it obtains an $\epsilon$-solution. Thus, this algorithm can be interpreted as an algorithm for solving an optimization problem with minimizing $\|Ax\|$ as its objective function. This interpretation makes the von Neumann algorithm to fall in the frame of the Frank–Wolfe algorithm [13]. Actually, when the Frank–Wolfe algorithm does the exact line search for calculating the step length at each iteration, the Frank–Wolfe algorithm exactly reduces to the von Neumann algorithm.

**Algorithm 2:** The von Neumann Algorithm

**Initialization**
Choose any $x \in \Delta_n$.
Let $b^0 = Ax^0$ and $k = 0$.
**While** $\|b^k\| \ge \epsilon$ **Do**

1. Find $a_s$ which makes the largest angle with the vector $b^k$, i.e., $a_s = Ax(b^k)$.
   Let $v_k = a_s^T b^k$.
2. **If** $v_k > 0$, STOP, problem (2) is infeasible.
3. Let $e_s$ be the unit vector corresponding to index $s$. Let

$$\lambda = \frac{1 - v_k}{\|b^k\|^2 - 2v_k + 1},$$

$$x^{k+1} = \lambda x^k + (1 - \lambda)e_s,$$

$$b^{k+1} = Ax^{k+1},$$

$$k = k + 1.$$

**End While**

In Algorithm 2, the vector $b$ is also called the residual at the current iterate. Dantzig assumed that problem (2) is feasible and derived an upper bound for the computational complexity [9] as follows.

**Theorem 2 ([9]).** *Let $\epsilon > 0$ and assume that problem (2) is feasible. Then the von Neumann algorithm provides an $\epsilon$-solution for problem (2) in at most*

$$\left\lceil \frac{1}{\epsilon^2} \right\rceil$$

*iterations.*

Epelman and Freund [11] gave a different complexity analysis and showed that when problem (2) is strictly feasible or strictly infeasible, then the iteration complexity of the von Neumann algorithm is linear in $\log(1/\epsilon)$ and $1/\rho_v(A)^2$.

**Theorem 3 ([11]).** *Suppose that $\rho_v(A) > 0$ and let $\epsilon > 0$.*

*(1) If problem (2) is strictly feasible, then in at most*

$$\left\lceil \frac{2}{\rho_v(A)^2} \ln \frac{1}{\epsilon} \right\rceil$$

*iterations, the von Neumann algorithm obtains an $\epsilon$-solution of problem (2).*
*(2) If problem (2) is strictly infeasible, then in at most*

$$\left\lfloor \frac{1}{\rho_v(A)^2} \right\rfloor$$

*iterations a certificate of infeasibility is given by the von Neumann algorithm.*

When problem (2) is feasible, both the complexity bounds proved by Dantzig (Theorem 2) and Epelman and Freund (Theorem 3) are for obtaining an $\epsilon$-approximate solution. Neither of them is dominant. When $\rho_v(A)$ is large, the complexity bound proved by Epelman and Freund is better. Otherwise, the one by Dantzig is better.

Theoretically the von Neumann algorithm does not provide an exact solution; it only converges to a solution. Dantzig [8] proposed a "bracketing" procedure to identify an exact solution in finite number of iterations. By applying the von Neumann algorithm to $m + 1$ perturbed problems, $m + 1$ approximate solutions are generated. A weighted sum of these approximate solutions yields an exact solution to the original unperturbed problem. This requires the solution of an $(m + 1) \times (m + 1)$ system of linear equations. This procedure has the following complexity.

**Theorem 4 ([8]).** *Suppose that problem (2) is strictly feasible. By applying the Dantzig's "bracketing" procedure, an exact feasible solution is found in*

$$\frac{4(m+1)^3}{\rho_v(A)^2}$$

*iterations.*

## 3  Duality

In this section, we first employ the Farkas Lemma to analyze the duality relationship between problem (1) and problem (2). This observation provides the foundation for the duality between the perceptron and the von Neumann algorithms. Then we extend the definition of $\rho$ to infeasible problems and give $\rho$ meaningful explanations for different problems. At last, we propose to utilize an Approximate Farkas Lemma to interpret an approximate solution from its dual perspective.

### 3.1  Alternative Systems

Recall that conv($A$) represents the convex hull of the points $a_i$. Assume that problem (1) is feasible and $y$ is a feasible solution. In this case, the hyperplane with normal vector $y$ separates conv($A$) from the origin, what implies that problem (2) is infeasible. Conversely, if problem (2) is infeasible, then there exists at least one separating hyperplane that can separate conv($A$) from the origin. In other words, there exists a vector $y$ such that $A^T y > 0$, which means problem (1) is feasible. Therefore, problem (1) and problem (2) are a pair of alternative systems. This conclusion can also be verified by the Farkas Lemma [3, 24, 26]. According to the Farkas Lemma, the alternative system of problem (2) is

$$A^T y + e\eta \geq 0,$$
$$\eta < 0. \tag{3}$$

 Problem (3) can equivalently be written as $A^T y > 0$, which is the form of problem (1). Thus, problems (1) and (2) are alternative systems to each other, i.e., exactly one of them is solvable. Since the perceptron and the von Neumann algorithms solve problems (1) and (2), respectively, the duality relationship between these two problems leads to a duality between the two algorithms.

### 3.2  Calculation of $\rho$

Section 2 gives a general definition of the condition number $\rho$. In this section, we discuss its special forms for the different problem forms in the different algorithms. The Classical Perceptron Algorithm shown in Sect. 2.1 assumes that problem (1) is strictly feasible. Thus, $\rho$ is only defined for feasible problems in [10]. In order to make the discussions about the duality complete, we extend the definition of $\rho$ to infeasible cases.

**Definition 3.** Consider the LFP (1).

1. If problem (1) is feasible [10], then the condition number $\rho_p(A)$ is the radius of the largest inscribed ball that fits in the feasible region, and the center of the ball is on the unit sphere. It is calculated by

$$\rho_p(A) = \max_{\|y\|=1} \min_i \{a_i^T y\}. \tag{4}$$

2. If problem (1) is infeasible, then $\rho_p(A)$ is the distance to feasibility, i.e.,

$$\rho_p(A) = \min_{\|y\|=1} \max_i \{-a_i^T y\}. \tag{5}$$

Note that when problem (1) is feasible, $\rho_p(A) > 0$ if and only if it is strictly feasible. On the other hand, the specific $\rho$ for problem (2) in the von Neumann algorithm is defined as follows.

**Definition 4 ([11]).** The condition number $\rho$ is the distance from the origin to the boundary $\partial(\text{conv}(A))$ of the feasible set $\text{conv}(A)$, i.e.,

$$\rho_v(A) = \inf\{\|h\| : h \in \partial(\text{conv}(A))\}. \tag{6}$$

Definition 4 also defines condition number $\rho_v(A)$ with two different meanings depending on the feasibility or infeasibility of problem (2).

1. If problem (2) is feasible, then $\rho_v(A)$ is the radius of the largest inscribed ball in $\text{conv}(A)$ centered at the origin. It can be calculated [20] by

$$\rho_v(A) = \min_{\|y\|=1} \max_i \{-a_i^T y\}. \tag{7}$$

When problem (2) is feasible but not strictly feasible, i.e., the origin is on the boundary of $\text{conv}(A)$, then $\rho_v(A) = 0$.
2. If problem (2) is infeasible, then $\rho_v(A)$ is the distance from the origin to $\text{conv}(A)$, i.e., $\rho_v(A)$ is the radius of the largest separating ball centered at the origin. It can be calculated as

$$\rho_v(A) = \max_{\|y\|=1} \min_i \{a_i^T y\}. \tag{8}$$

Comparing (4), (5), (7), and (8), it is easy to see that when problem (2) is infeasible (feasible), the condition number $\rho$ is computed in the same way as the one when problem (1) is feasible (infeasible). This observation originates from the alternative systems relationship of these two problems.

## 3.3 Interpret Approximate Solutions

Instead of providing an exact feasible solution as the perceptron algorithm does, the von Neumann algorithm returns an $\epsilon$-solution when it terminates in a finite number of iterations. Analogously, the Modified Perceptron Algorithm [5]—a variant of the perceptron algorithm—returns a $\sigma$-feasible solution when the perceptron problem (1) is close to the boundary of feasibility and infeasibility. A $\sigma$-feasible solution is also an approximate solution which we will discuss later. When $\epsilon$ or $\sigma$ is a fixed number, an $\epsilon$-solution or a $\sigma$-feasible solution is not sufficient to draw a firm conclusion about feasibility of the problem. In this section and also in the following section, our goal is to give some interpretations of these approximate solutions from their alternative perspective and answer the following questions:

- How to derive meaningful information from these approximate solutions?
- What conclusion can be drawn about the feasibility status of the problems?

The duality relationship discussed in Sect. 3.1 is directly derived from the Farkas Lemma. The two problems (1) and (2) are a pair of alternative systems and therefore, exactly one of them is solvable. Recall that both the Classical Perceptron and the von Neumann algorithms are non-polynomial algorithms. When the problems are close to the boundary of feasibility and infeasibility, both the classical perceptron and von Neumann algorithms are inefficient. It takes exponentially many iterations for the algorithms to obtain a clear answer about solvability of the problems. Therefore, we would like that some variants of the algorithms could provide an approximate solution or some indications of approximate feasibility or infeasibility. Due to the alternative system relationship between problems (1) and (2), a proof of infeasibility for one problem can be given by giving a solution to the other one. We are interested in exploring approximate solutions to this pair of alternative systems and their interpretations for their alternative systems. We first discuss the $\sigma$-feasible solution. It is defined as follows [5].

**Definition 5.** A vector $y$ is a $\sigma$-feasible solution (or $\sigma$-solution for short reference) to problem (1) if $A^T \bar{y} \geq -\sigma e$, where $\sigma$ is a small positive number.

According to the Definition 5, a $\sigma$-solution allows slight violations to the constraints in problem (1); and thus it is an approximate solution. Recall that we analyze the meaning of the condition number $\rho$ in Sect. 3.2. From Definition 4, we obtain the following theorem.

**Theorem 5.** *The following two statements are equivalent:*

*(a) The perceptron problem (1) has a $\sigma$-solution.*
*(b) There is no ball in* conv($A$) *centered at the origin and its radius is larger than $\sigma$.*

This theorem is straight derived from (7). Theorem 5 shows that a $\sigma$-feasible solution to the perceptron problem (1) indicates that the corresponding von Neumann problem (2) is either infeasible, or if it is feasible then it is close to infeasibility. As a result, we define such a $\sigma$-solution as a $\sigma$-infeasibility certificate for the von Neumann problem (2).

**Definition 6.** A vector $y$ is a $\sigma$-infeasibility certificate for the von Neumann problem (2) if $A^T \tilde{y} \geq -\sigma e$.

Combine Theorem 5 and Definition 6, we derive the following corollary.

**Corollary 1.** *A $\sigma$-infeasibility certificate indicates that the von Neumann problem (2) is either infeasible or feasible with $\rho_v(A) \leq \sigma$.*

## 3.4  Approximate Farkas Lemma

In the previous section, we interpret a $\sigma$-feasible solution to the perceptron problem (1) as a $\sigma$-infeasibility certificate of the von Neumann problem (2). In this section, we explore whether we can obtain an analogous result about an $\epsilon$-solution. The major tool we employ is the Approximate Farkas Lemma [30].

The Approximate Farkas Lemma is derived by Todd and Ye [30] from the general gauge duality results of Freund [14]. These lemmas are extensions of the Farkas Lemma [26] and quantify how certain approximate feasible solutions to a system of inequalities indicate the infeasibility of its alternative system. In order to adapt the Approximate Farkas Lemma, we first transfer a strictly feasible problem (1) to an optimization problem. Consider the following optimization problem:

$$\alpha_{\tilde{y}} = \min \|\tilde{y}\| \\ \text{s.t. } A^T \tilde{y} \geq e, \tag{9}$$

where $\tilde{y} \in \mathbb{R}^m$, matrix $A$ has the same definition as in problem (1), and $\alpha_{\tilde{y}}$ denote the optimal value. This optimization problem is to find a feasible solution to inequality system $A^T \tilde{y} \geq e$ with the shortest length. Comparing problem (9) and problem (1), any feasible solution to problem (9) is also a feasible solution to problem (1). On the other hand, if $y^*$ is a strictly feasible solution to problem (1), i.e., all coordinates of $A^T y^* > 0$, then $\tilde{y}^* = \frac{y^*}{(y^*)^T Ax(y^*)}$ is a feasible solution to problem (9), where $(y^*)^T Ax(y^*)$ gives the smallest coordinate of $A^T y^*$. Thus, feasibility of problem (9) is equivalent to strictly feasibility of problem (1). When problem (1) is strictly feasible, we can put a ball into the feasible region of $A^T y > 0$. If the radius of the ball is fixed to 1, then $\alpha_{\tilde{y}}$ measures the minimal distance from the center of this unit ball to the origin. Recall that $\rho_p(A)$ measures the radius of the maximal ball put in the feasible region and centered on the unit sphere. Comparing $\alpha_{\tilde{y}}$ and $\rho_p(A)$, the closer problem (1) is to infeasibility, the narrower the feasible region is, the smaller $\rho_p(A)$ is, and the further the unit ball is from the origin, i.e., the larger $\alpha_{\tilde{y}}$ is. Thus, $\alpha_{\tilde{y}}$ is seen as another measure of the robustness of problem (1). We obtain the following result by geometrical relationship:

$$\alpha_{\tilde{y}} = \frac{1}{\rho_p(A)}. \tag{10}$$

By adapting the Approximate Farkas Lemma [30] to our problem, we obtain:

**Lemma 1 (Approximate Farkas Lemma).** *Consider the optimization problems*

$$(GP): \quad \alpha_{\bar{y}} = \min\left\{ \|\tilde{y}\| \mid A^T \tilde{y} \geq e \right\},$$

*and*

$$(GD): \quad \beta_b = \min\left\{ \|b\| \mid Ax = b, e^T x = 1, x \geq 0 \right\}.$$

*Then* $\alpha_{\bar{y}} \beta_b = 1$.

The special case $0 \cdot +\infty$ is interpreted as 1. It is easy to see that problem (GD) is the perturbed problem of problem (2). When problem (1) is strictly feasible, $\beta_b$ gives the minimal distance between the origin and conv($A$), which is the radius $\rho_v(A)$ of the largest separating ball defined by (8). $\beta_b$ also indicates the minimum corrections needed to make problem (2) feasible. By Lemma 1 and (10), we have $\beta_b = \frac{1}{\alpha_{\bar{y}}} = \rho_p(A)$. Therefore, the Approximate Farkas Lemma verifies that $\rho_p(A)$ for feasible perceptron problem (1) is equivalent to $\rho_v(A)$ for infeasible von Neumann problem (2). Thus, any feasible solution to problem (GP) is an infeasibility certificate of problem (2) and gives a lower bound for the distance to feasibility.

On the other side, if problem (2) is feasible, then its any feasible solution is an optimal solution to optimization problem (GD) with $\beta_b = 0$. According to the Approximate Farkas Lemma, $\alpha_{\bar{y}} = +\infty$. It implies that problem (GP) is infeasible, then Lemma 1 reduces to the exact Farkas Lemma. In this case, problem (1) is either infeasible or feasible but not strictly feasible.

Now we have the following theorem which utilizes the Approximate Farkas Lemma to interpret an approximate solution.

**Theorem 6.** *The following three statements are equivalent:*

*(a)* *The von Neumann problem (2) has an $\epsilon$-solution.*
*(b)* *A unit ball cannot be put closer than $1/\epsilon$ to the origin in the feasible region of problem (1).*
*(c)* *There is no ball in the feasible region of problem (1) centered on the unit sphere and its radius is larger than $\epsilon$.*

*Proof.* Problem (2) has an $\epsilon$-solution $x'$ such that $\beta_b' = \|b'\| = \|Ax'\| \leq \epsilon$, if and only if $\beta_b \leq \beta_b' \leq \epsilon$. By Lemma 1, this holds if and only if $\frac{1}{\epsilon} \leq \frac{1}{\beta_b} = \alpha_{\bar{y}}$. The inequality $\frac{1}{\epsilon} \leq \alpha_{\bar{y}}$ holds if and only if

$$\nexists \tilde{y} \text{ such that } \|\tilde{y}\| < \frac{1}{\epsilon} \text{ and } A^T \tilde{y} \geq e. \tag{11}$$

By scaling, (11) is equivalent to

$$\nexists y \text{ such that } \|y\| \leq 1 \text{ and } A^T y > \epsilon e. \tag{12}$$

Statement (11) is the statement (b). Recall the definition of $\rho_p(A)$; (12) indicates that $\rho_p(A) \leq \epsilon$. Thus, statements (a), (b), and (c) are equivalent.

Theorem 6 shows that an $\epsilon$-solution to the von Neumann problem (2) implies that the corresponding perceptron problem (1) is either infeasible, or if it is feasible then it is close to infeasibility. Therefore, we define such an $\epsilon$-solution as an $\epsilon$-infeasibility certificate for problem (1).

**Definition 7.** A vector $y$ is an $\epsilon$-infeasibility certificate for the perceptron problem (1) if there exists a vector $x \in \Delta_n$ such that $Ax = y$ and $\|y\| \leq \epsilon$.

We can derive the following corollary from Theorem 6.

**Corollary 2.** *An $\epsilon$-infeasibility certificate indicates that the perceptron problem (1) is either infeasible or feasible with $\rho_p(A) \leq \epsilon$.*

Since $\epsilon$ is a small positive number, it means that the norm of any feasible solution $\tilde{y}'$ of problem (GP), if it exists, is at least as large as $\frac{1}{\epsilon}$. Thus, if problem (GP) is close to infeasibility, its feasible solutions have to be large. For example, if problem (GP) is feasible and the distance to the infeasibility is as small as $10^{-10}$, then the magnitude of a feasible solution $\tilde{y}'$ has to be at least $10^{10}$ large.

We utilized the definition of the condition number $\rho$ and the Approximate Farkas Lemma to interpret approximate solutions so that we can draw more definitive conclusions about the solutions or feasibility of the problems. In addition, when the respective variants of the perceptron and the von Neumann algorithms terminate at a certain point, then the Approximate Farkas Lemma allows a more precise interpretation of the output and provides meaningful information about the solution.

Inspired by the alternative system relationship of problems (1) and (2), we investigate the duality of the perceptron and the von Neumann algorithms. In Sect. 4, different versions of the perceptron algorithm are interpreted as variants of the von Neumann algorithm as they are applied to problem (2). In Sect. 5, we interpret variants of the von Neumann algorithm from the perspective of the perceptron algorithm. By exploring this intriguing duality of these algorithms, we not only gain new insight into the intimate relationship of these algorithms but also derive several novel variants of these algorithms with their corresponding complexity results.

## 4  From Perceptron to von Neumann

Since problems (1) and (2) are alternative systems to each other, the perceptron algorithm can be operated on problem (2) with proper adjustments. The complexity results for the feasible case of the perceptron algorithm are adaptable for the infeasible case of the von Neumann algorithm. Since the perceptron algorithm has several variants, we discuss them in the following subsections.

In order to make our discussions more transparent, we rename the two spaces. The perceptron algorithm solves problems in form (1) to get a solution $y$ if the problem is feasible. Thus, we call the space $\mathbb{R}^m$ in which the vector $y$ lives the *perceptron space*. Similarly, because the von Neumann algorithm solves problem (2), we call $\mathbb{R}^n$ the *von Neumann space*. Note that the vector $b^k = Ax^k$ in the von Neumann algorithm, presented as Algorithm 2, is in the perceptron space. This reflects the duality of the two problems and also indicates some close relationships between the two algorithms. Matrix $A$ can be seen as a linear operator between the perceptron and the von Neumann spaces.

## 4.1   The Normalized Perceptron Algorithm

We first revisit, Algorithm 1, the Classical Perceptron Algorithm. It starts from $y^0 = 0$ and at iteration $k$, from the point $y^k$ it makes one unit step in the direction of a violated constraint $a_j$. Let $x^k$ be the corresponding vector that satisfies $y^k = Ax^k$. We have $x^0 = 0$ and $x^{k+1} = x^k + e_j$, where $Ae_j = a_j$. According to this relationship, $x^k$ is a sequence of vectors in the von Neumann space with $x^k \geq 0$ and $\|x^k\|_1 = k$ for all $k \in \mathbb{N}$. On the other hand, observe that the last two constraints in problem (2), $e^T x = 1, x \geq 0$ restrict vector $x$ to be in the unit simplex $\Delta_n$. The comparison of $x^k$ at iteration $k$ in the Classical Perceptron Algorithm and $x$ in problem (2) leads to a normalized version of the perceptron algorithm [29]. Assume that problem (1) is feasible. The Normalized Perceptron Algorithm is presented as Algorithm 3.

Note that at the end of each iteration, the iterate $y^k$ is inspected. The algorithm terminates if $y^k$ is an $\epsilon$-infeasibility certificate. This stopping criterion is derived from the von Neumann side after we successfully explain an approximate solution. In the Normalized Perceptron Algorithm, the $k$th iterate $y^k$ is divided by $k$ to satisfy $y^k = Ax^k$ for some $x^k \in \Delta_n$. Thus, $x^k$ is a vector $x$ in the von Neumann space, and $y^k$ can be interpreted as the corresponding $b^k$ vector in the von Neumann algorithm. If the Normalized Perceptron Algorithm starts from $x^0 = 0$ and $x^k$ can be updated to satisfy $x^k \in \Delta_n$ and $y^k = Ax^k$, then we get a variant of the von Neumann algorithm. To ease understanding, the derived Normalized von Neumann Algorithm is described in full details in Algorithm 4.

When applying to problem (2), the Normalized von Neumann Algorithm has the following complexity result.

**Theorem 7.** *Let $\epsilon > 0$.*

*(1) If problem (2) is feasible, then the Normalized von Neumann Algorithm provides an $\epsilon$-solution in at most*

$$\left\lceil \frac{1}{\epsilon^2} \right\rceil$$

*iterations.*

**Algorithm 3:** The Normalized Perceptron Algorithm

**Initialization** Let $y^0 = 0$ and $k = 0$.
**While True Do**
Find a column $a_j$, $j \in \{1, 2, \cdots, n\}$ such that $a_j^T y^k \le 0$.
**If** such an $a_j$ does not exist, STOP and return $y^k$.
   **Else**

$$\theta_k = \frac{1}{k+1},$$

$$y^{k+1} = (1 - \theta_k)y^k + \theta_k a_j,$$

$$k = k + 1.$$

**If** $\|y^k\| \le \epsilon$, STOP and return $y^k$ as an $\epsilon$-infeasibility certificate.
**End While**

(2) *If problem (2) is strictly infeasible, then in at most*

$$\left\lceil \frac{1}{\rho_v(A)^2} \right\rceil$$

*iterations an infeasibility certificate is given.*

*Proof.* When problem (2) is strictly infeasible, then its alternative problem (1) is strictly feasible. Since $y^k$, generated by the Normalized Perceptron Algorithm, is exactly the same as $y^k$ in the Classical Perceptron Algorithm divided by $k$, the complexity result of the Classical Perceptron Algorithm stated in Theorem 1 is also valid for Algorithm 3, the Normalized Perceptron Algorithm. Thus, the complexity result for strictly infeasible problems is an immediate corollary of Theorem 1.

Now we prove the complexity when problem (2) is feasible. By using induction on $k$, we prove that $\|b^k\| \le \frac{1}{\sqrt{k}}$. For $k = 1$, since the algorithm starts with $b^0 = 0$,

$$\|b^1\| = \|(1 - \theta_0)b^0 + \theta_0 Ax(b^0)\| = \|Ax(b^0)\| = 1,$$

where the last equality results from $\|a_i\| = 1$ for $i = 1, \ldots, n$.
Now, suppose that we have $\|b^{k-1}\| \le \frac{1}{\sqrt{k-1}}$. At the iteration $k$, we obtain

$$\|b^k\|^2 = \|(1 - \theta_{k-1})b^{k-1} + \theta_{k-1}a_j\|^2$$

$$= (1 - \theta_{k-1})^2\|b^{k-1}\|^2 + \theta_{k-1}^2\|a_j\|^2 + 2\theta_{k-1}(1 - \theta_{k-1})(a_j^T b^{k-1})$$

$$\le \frac{1}{k^2}\left[(k-1)^2\|b^{k-1}\|^2 + 1\right] \le \frac{1}{k}$$

The first inequality must be true because $a_j^T b^{k-1} \le 0$ when problem (2) is feasible. The second inequality holds due to the inductive hypothesis $\|b^k\| \le 1/\sqrt{k}$. Thus,

$$\epsilon = \|b^k\| \le 1/\sqrt{k}.$$

Therefore, the algorithm needs at most $\lceil 1/\epsilon^2 \rceil$ iterations.                    $\square$

## Algorithm 4: The Normalized von Neumann Algorithm

**Initialization** Let $x^0 = 0$, $b^0 = Ax^0$, and $k = 0$.
**While** $\|b^k\| \geq \epsilon$ **Do**
Find an $a_j$, $j \in \{1, 2, \cdots, n\}$ such that $a_j^T b^k \leq 0$.
**If** such an $a_j$ does not exist, STOP, return $b^k$ as an infeasibility certificate of problem (2).
　　**Else**

$$\theta_k = \frac{1}{k+1},$$

$$x^{k+1} = (1 - \theta_k)x^k + \theta_k a_j,$$

$$b^{k+1} = Ax^{k+1},$$

$$k = k + 1.$$

**End While**
**Return** $x^k$ as an $\epsilon$-solution.

When problem is feasible, Theorem 7 shows that the complexity result of the Normalized von Neumann Algorithm is the same as Dantzig's result (Theorem 2) for the von Neumann algorithm. However, there are two major differences between the Normalized von Neumann Algorithm and the original von Neumann algorithm. At each iteration, the original von Neumann algorithm searches $a_s$ which has the largest angle with $b^k$ and computes step-length $\lambda$ based on the current iterate $b^k$ and $a_s$. The Normalized von Neumann Algorithm, which is transformed from the Classical Perceptron Algorithm, uses any $a_j$ which satisfies $a_j^T b^k \leq 0$. Its update step length only depends on $k$. Thus, the cost per iteration of the von Neumann algorithm is $2n^2$ more than that of the Normalized von Neumann Algorithm.

Recall that Theorem 1 provides the complexity result for feasible perceptron problems. If problem (1) is strictly feasible, then the Classical Perceptron Algorithm returns a feasible solution in at most $1/\rho_p(A)^2$ iterations. However, there is no published result for infeasible perceptron problems. Now by transiting Theorem 7 back to the perceptron problems, we obtain the following new result for the Classical Perceptron Algorithm.

**Theorem 8.** *Let $\epsilon > 0$. If problem (1) is infeasible, then the Classical/Normalized Perceptron Algorithm provides an $\epsilon$-infeasibility certificate in at most*

$$\left\lceil \frac{1}{\epsilon^2} \right\rceil$$

*iterations, which indicates that there is no such an $\epsilon$-ball in the feasible region.*

The complexity bound in Theorem 8 only depends on the value $\epsilon$, the accuracy of the infeasibility certificate, but does not depend on the geometry of the problem.

## 4.2   The Smooth Perceptron Algorithm

Soheili and Peña [29] proposed a smooth version of the perceptron algorithm and showed that it can be seen as a smooth first-order algorithm. This deterministic variant retains the original simplicity of the perceptron algorithm and its complexity is improved by almost a factor of $1/\rho_p(A)$ compared to the Classical Perceptron Algorithm. The improved complexity result is given in Theorem 9. We first introduce the Smooth Perceptron Algorithm.

Given $\mu > 0$, $x(y)$ is smoothed by the entropy prox-function

$$x_\mu(y) = \frac{e^{\frac{-A^T y}{\mu}}}{\left\| e^{\frac{-A^T y}{\mu}} \right\|_1}, \tag{13}$$

where the expression $e^{\frac{-A^T y}{\mu}}$ denotes the $n$-dimensional vector

$$e^{\frac{-A^T y}{\mu}} = \left[ e^{\frac{-a_1^T y}{\mu}}, e^{\frac{-a_2^T y}{\mu}}, \ldots, e^{\frac{-a_n^T y}{\mu}} \right]^T.$$

The Smooth Perceptron Algorithm is presented in Algorithm 5.

Compared to the complexity of the Classical Perceptron Algorithm stated in Theorem 1, Theorem 9 shows that the Smooth Perceptron Algorithm has a complexity result with $\frac{1}{\rho_p(A)\sqrt{\log(n)}}$ improvement.

**Theorem 9 ([29]).** *Assume that problem (1) is strictly feasible. The Smooth Perceptron Algorithm terminates in at most*

$$\frac{2\sqrt{\log(n)}}{\rho_p(A)} - 1$$

*iterations with a feasible solution.*

Analogous to the Normalized Perceptron Algorithm, the Smooth Perceptron Algorithm can also be applied to problem (2) when it is infeasible (see Algorithm 6). Iterate $y^k$ in the perceptron space plays the role of the vector $b^k$ in the von Neumann algorithm. Since $b^k$ is updated so that $Ax^k = b^k$, we derive the corresponding vector $x^k$.

Compare Algorithm 5 and 6. Iterate $y^k$ in Algorithm 5 is the same as vector $b^k$ in Algorithm 6, and its corresponding $x^k$ satisfying $y^k = Ax^k$ is the vector $x$ in problem (2). It is easy to see that $x^k \in \Delta_n$ for all iterations. Therefore, the complexity result of Theorem 9 applies to Algorithm 6 as well when problem (2) is infeasible. We derive the following corollary from Theorem 9.

**Corollary 3.** *Assume that problem (2) is strictly infeasible. The Smooth von Neumann Algorithm terminates in at most*

**Algorithm 5:** The Smooth Perceptron Algorithm

**Initialization** Let $y^0 = \frac{Ae}{n}$, $\mu_0 = 1$, and $x^0 = x_{\mu_0}(y^0)$. $k = 0$.
**While True Do**
Let $a_s = Ax(y^k)$.
**If** $a_s^T y^k > 0$, STOP and return $y^k$.
   **Else**

$$\theta_k = \frac{2}{k+3},$$

$$y^{k+1} = (1 - \theta_k)(y^k + \theta_k Ax^k) + \theta_k^2 Ax_{\mu_k}(y^k),$$

$$\mu_{k+1} = (1 - \theta_k)\mu_k,$$

$$x^{k+1} = (1 - \theta_k)x^k + \theta_k x_{\mu_{k+1}}(y^{k+1}),$$

$$k = k + 1.$$

**End While**

**Algorithm 6:** The Smooth von Neumann Algorithm

**Initialization** Let $x^0 = \frac{e}{n}$, $b^0 = Ax^0 = \frac{Ae}{n}$, $\mu_0 = 1$, and $\tilde{x}^0 = x_{\mu_0}(y^0)$. $k = 0$.
**While True Do**
Let $a_s = Ax(b^k)$
**If** $a_s^T b^k > 0$, STOP and return $b^k$ as an infeasibility certificate.
   **Else**

$$\theta_k = \frac{2}{k+3},$$

$$b^{k+1} = (1 - \theta_k)(b^k + \theta_k A\tilde{x}^k) + \theta_k^2 Ax_{\mu_k}(b^k),$$

$$\mu_{k+1} = (1 - \theta_k)\mu_k,$$

$$\tilde{x}^{k+1} = (1 - \theta_k)\tilde{x}^k + \theta_k x_{\mu_{k+1}}(b^{k+1}),$$

$$x^{k+1} = (1 - \theta_k)(x^k + \theta_k \tilde{x}^k) + \theta_k^2 x_{\mu_k}(b^k),$$

$$k = k + 1.$$

**End While**

$$\frac{2\sqrt{\log(n)}}{\rho_v(A)} - 1$$

*iterations with a certificate of infeasibility $b^k$ such that $A^T b^k > 0$.*

Recall that the Smooth Perceptron Algorithm has an improved complexity result compared to the Normalized Perceptron Algorithm when problem (1) is feasible. Thus, if problem (2) is infeasible, then after being interpreted in the von Neumann space, the Smooth von Neumann Algorithm also enjoys an almost $1/\rho_v(A)$ complexity improvement compared to the one presented in Corollary 7.

Independently of our work, Soheili and Peña [28] proposed a version of smooth von Neumann algorithm called Iterated Smooth Perceptron-von Neumann (ISPVN) Algorithm. It is also based on the duality relationship between problems (1) and (2). Instead of using the entropy prox-function as the Smooth Perceptron Algorithm, it employs the Euclidean prox-function to smooth $x(y)$. The merit of the ISPVN Algorithm is that when problem (1) is infeasible with $\rho_p(A) > 0$, the ISPVN Algorithm solves its alternative system (2). Thus, the ISPVN Algorithm could handle both problems (1) and (2) simultaneously. It either finds a feasible solution to problem (1) in $O\left(\frac{\sqrt{n}}{\rho_p(A)}\log\left(\frac{1}{\rho_p(A)}\right)\right)$ elementary iterations or finds an $\epsilon$-solution to the corresponding problem (2) in $O\left(\frac{\sqrt{n}}{\rho_v(A)}\log\left(\frac{1}{\epsilon}\right)\right)$ elementary iterations. Both of the iteration complexity of the ISPVN Algorithm are better than these of the perceptron and the von Neumann algorithms. However, in the case when problem (2) is infeasible, the Smooth von Neumann Algorithm stated in Algorithm 6 has a better complexity bound.

## 5 From von Neumann to Perceptron

After interpreting variants of the perceptron algorithm from its dual perspective in Sect. 4, in this section we show how to interpret the von Neumann algorithm as a variant of the perceptron algorithm and how to apply it to problem (1).

### 5.1 The Original von Neumann Algorithm

The von Neumann algorithm was reviewed in Sect. 2.2. Note that in the von Neumann algorithm, iterates $x^k$ are always in the unit simplex. The vector $b^k = Ax^k$ in the von Neumann algorithm can play the role of vector $y$ in the perceptron space. According to Theorem 3, there are two possible outcomes of the von Neumann algorithm. If problem (2) is strictly infeasible, the von Neumann algorithm provides an infeasibility certificate. Then the alternative case in the perceptron space is that problem (1) is strictly feasible. Thus, applying the von Neumann algorithm to problem (1) will provide a feasible solution $y^k$. On the other hand, if problem (2) is strictly feasible, then the von Neumann algorithm will return an $\epsilon$-solution with $\|b^k\| < \epsilon$, and $m+1$ applications of the von Neumann algorithm allow to get an exact solution [8], which is interpreted as an exact infeasibility certificate for problem (1). However, if problem (2) is neither strictly feasible with an $\epsilon$-ball in the feasible set nor strictly infeasible with at least $\epsilon$ distance to feasibility, then an $\epsilon$-solution interpreted in the perceptron space implies an $\epsilon$-infeasibility certificate of problem (1). An $\epsilon$-solution/$\epsilon$-infeasibility certificate could have two possible meanings.

**Algorithm 7:** The von Neumann algorithm in the perceptron space

**Initialization**
Choose any $x^0 \in \Delta_n$. Let $y^0 = Ax^0$ and $k = 0$.
**While** $\|y^k\| \geq \epsilon$ **Do**

1. Let $\mu_k = \|y^k\|$ and $v_k = (Ax(y^k))^T y^k$, where $x(y^k) = \underset{x \in \Delta_n}{\operatorname{argmin}} \{(y^k)^T Ax\}$.

2. If $v_k > 0$, STOP and return $y^k$ as a solution.
3. Update

$$\lambda = \frac{1 - v_k}{\mu_k^2 - 2v_k + 1},$$

$$\mu_{k+1}^2 = \lambda v_k + (1 - \lambda),$$

$$y^{k+1} = \lambda y^k + (1 - \lambda)Ax(y^k),$$

$$k = k + 1.$$

**End While**

1. If problem (2) is feasible, then problem (1) is infeasible.
2. If problem (2) is infeasible, then an $\epsilon$-solution implies that the distance to the infeasibility of problem (2) is less than $\epsilon$, i.e., $\rho_v(A) < \epsilon$; and the radius of the largest inscribed ball in the feasible region of problem (1) is $\rho_p(A) < \epsilon$. This means that though problem (1) is feasible, it is almost infeasible. The distance to the infeasibility is less than $\epsilon$.

Thus, problem (1) is either infeasible or $\epsilon$-close to infeasibility. From Theorem 3 the following complexity result can be derived for Algorithm 7.

**Theorem 10.** *Let $\epsilon > 0$.*

*(1) If problem (1) is strictly feasible, then in at most*

$$\left\lfloor \frac{1}{\rho_p^2} \right\rfloor$$

*iterations the von Neumann algorithm finds a feasible solution to problem (1).*
*(2) If problem (1) is strictly infeasible, then in at most*

$$\left\lceil \frac{2}{\rho_p^2} \ln \frac{1}{\epsilon} \right\rceil$$

*iterations the von Neumann algorithm provides an $\epsilon$-infeasibility certificate.*

## 5.2   The Optimal Pair Adjustment Algorithm

Gonçalves et al. [17] introduced three variants of the von Neumann algorithm named as Weight-Reduction, Optimal Pair Adjustment (OPA), and Projection Algorithms. Among these three variants, the OPA Algorithm has the best performance in computational experiments. The basic idea of the OPA Algorithm is to move the residual $b^k$ in Algorithm 2 closer to the origin 0 as much as possible at each update step. It gives the maximum possible freedom to two weights: the one in column $a_{s+}$ which has the largest angle with $b^k$ and the one in column $a_{s-}$ which has the smallest angle with $b^k$. At each iteration, it finds the optimal value for these two coordinates and adjusts the remaining ones. The OPA Algorithm is presented as Algorithm 8.

**Algorithm 8:** The Optimal Pair Adjustment Algorithm

**Initialization**
Choose any $x^0 \in \Delta_n$. Let $b^0 = Ax^0$.
Given a small positive number $\epsilon$.
**While $\|b^k\| \geq \epsilon$ Do**

1. **Find** the vectors $a_{s+}$ and $a_{s-}$ which make the largest and smallest angles with the current iterate $y^k$:

$$s^+ = \underset{i=1,\dots,n}{\mathrm{argmin}} \{a_i^T b^k\},$$

$$s^- = \underset{i=1,\dots,n}{\mathrm{argmin}} \{a_i^T b^k \mid x_i > 0\},$$

$$v_k = a_{s+}^T b^k.$$

2. **If** $v_k > 0$, STOP, return $b^k$ as a feasible solution to problem (2).
3. **Solve** the subproblem

$$\min \qquad \|\bar{b}\|^2$$
$$\text{s.t. } \lambda_0 (1 - x_{s+}^k - x_{s-}^k) + \lambda_1 + \lambda_2 = 1,$$
$$\lambda_i \geq 0, \text{ for } i = 1, 2,$$

where $\bar{b} = \lambda_0 (b^k - x_{s+}^k a_{s+} - x_{s-}^k a_{s-}) + \lambda_1 a_{s+} + \lambda_2 a_{s-}$.

4. **Update**

$$b^{k+1} = \lambda_0 (b^k - x_{s+}^k a_{s+} - x_{s-}^k a_{s-}) + \lambda_1 a_{s+} + \lambda_2 a_{s-},$$

$$x_i^{k+1} = \begin{cases} \lambda_0 x_i^k & i \neq s^+, s^-, \\ \lambda_1, & i = s^+, \\ \lambda_2, & i = s^-. \end{cases}$$

$$k = k + 1.$$

**End While**

The OPA Algorithm has a better performance than the original von Neumann algorithm in practice [17], and Gonçalves proves that in theory it is at least as good as the original von Neumann algorithm.

**Theorem 11 ([17]).** *The decrease in $\|b^k\|$ obtained by an iteration of the OPA Algorithm is at least as large as that obtained by one iteration of the von Neumann algorithm.*

Therefore, the OPA and the von Neumann algorithms share the same theoretical complexity as given in Theorem 3.

In the dual space, the residual $b^k$ is the normalized iterate $y^k$ in the perceptron algorithm. The column which has the largest angle with $b^k$ corresponds the "most infeasible" constraint for $y^k$. Since a feasible solution to the von Neumann problem is an infeasibility certificate for the corresponding perceptron problem, the faster the residual $b^k$ moves closer to 0, the sooner infeasibility of the perceptron problem is detected. Therefore, the OPA algorithm outperforms the von Neumann algorithm when problem (1) is infeasible. In this section, we describe the equivalent OPA Perceptron Algorithm in Algorithm 9.

The subproblem in Step 3 can be solved by enumerating all possibilities that satisfy the KKT conditions [17]. Analogous to Algorithm 7, if problem (1) is feasible and Algorithm 9 terminates with $u^k < \epsilon$, then there is no $\epsilon$-ball contained in the feasible cone centered on the unit sphere. In this case, problem (1) is close to infeasibility and $y^k$ is an $\epsilon$-infeasibility certificate. After interpreted as a variant of the perceptron algorithm, the OPA Perceptron Algorithm has the complexity result as stated in Theorem 10.

# 6   Summary and Future Work

The perceptron and the von Neumann algorithms are used to solve LFPs in different forms. In this paper, we reveal the duality relationship between these algorithms. This observation is based on the fact that the forms of the LFPs these algorithms deal with are a pair of Farkas alternative systems. This relationship enables us to interpret variants of the perceptron algorithm as variants of the von Neumann algorithm and vice versa. The dual interpretation of the algorithms allows us to transit the complexity results to the new algorithms too. The interpretation of an approximate solution is crucial during the algorithms transit. Utilizing the Approximate Farkas Lemma makes the solution meaningful for the alternative system and the transit complete. A major difference of these two algorithm families is that the perceptron algorithm assumes that the problem is feasible while the von Neumann algorithm solves both feasible and infeasible problems. Therefore, in this paper, we show that the infeasibility of perceptron problems are detected by the interpreted von Neumann algorithm (Algorithm 7) and the OPA Perceptron Algorithm (Algorithm 9); and the Normalized von Neumann Algorithm (Algorithm 4)—interpreted from the Normalized Perceptron Algorithm—is applicable for both strictly infeasible and feasible von Neumann problems. Furthermore, when problem (1) is infeasible,

**Algorithm 9:** The Optimal Pair Adjustment Perceptron Algorithm

**Initialization**
Choose any $x^0 \in \Delta_n$. Let $y^0 = Ax^0$ and $u^0 = \|y^0\|$. $\theta = 1$.
Given a small positive number $\epsilon$.
**While $u^k \geq \epsilon$ Do**

1. **Find** the vectors $a_{s+}$ and $a_{s-}$ which make the largest and smallest angles with the current iterate $y^k$:

$$s^+ = \operatorname*{argmin}_{i=1,\ldots,n} \{a_i^T y^k\},$$

$$s^- = \operatorname*{argmin}_{i=1,\ldots,n} \{a_i^T y^k | x_i > 0\},$$

$$v_k = a_{s+}^T y^k.$$

2. **If** $v_k > 0$, STOP. Return $y^k$ as a feasible solution to problem (1).
3. **Solve** the subproblem

$$\min \qquad \|\bar{y}\|^2$$
$$\text{s.t. } \lambda_0(1 - x_{s+}^k - x_{s-}^k) + \lambda_1 + \lambda_2 = 1,$$
$$\lambda_i \geq 0, \text{ for } i = 1, 2,$$

where $\bar{y} = \lambda_0(\frac{y^k}{\theta} - x_{s+}^k a_{s+} - x_{s-}^k a_{s-}) + \lambda_1 a_{s+} + \lambda_2 a_{s-}$.
4. **Update**
   **If $\lambda_0 = 0$, then**

$$\theta = 1,$$
$$y^{k+1} = \lambda_1 a_{s+} + \lambda_2 a_{s-}.$$

   **Else,**

$$\theta = \frac{\theta}{\lambda_0}$$
$$y^{k+1} = y^k + (\theta\lambda_1 - x_{s+}^k)a_{s+} + (\theta\lambda_2 - x_{s-}^k)a_{s-}.$$

   **End If**

$$u^{k+1} = \frac{1}{\theta}\|y^{k+1}\|,$$

$$x_i^{k+1} = \begin{cases} \lambda_0 x_i^k & i \neq s^+, s^-, \\ \lambda_1, & i = s^+, \\ \lambda_2, & i = s^-. \end{cases}$$

$$k = k + 1.$$

**End While**

we derive a complexity result for the Classical Perceptron Algorithm from the perspective of the von Neumann space.

There is another variant of the perceptron algorithm—the Modified Perceptron Algorithm [5]. It starts from a random vector $y$. In order to interpret it from the von Neumann perspective, finding the corresponding vector $x$ is a critical step. In addition, the Modified Perceptron Algorithm returns a $\sigma$-feasible solution, which is also an approximate solution—when the perceptron problem is feasible with small quantity $\rho_p(A)$—smaller than a given threshold. Therefore, we are aiming to design a modified version of von Neumann algorithm which can return an infeasibility certificate when the problem is almost infeasible. Its interpreted variant will benefit detecting infeasibility of perceptron problems when $\rho_p(A)$ is small.

# References

1. Bárány, I., Onn, S.: Colourful linear programming and its relatives. Mathe. Oper. Res. **22**(3), 550–567 (1997)
2. Beck, A., Teboulle, M.: A conditional gradient method with linear rate of convergence for solving convex linear systems. Math. Methods Oper. Res. **59**, 235–247 (2004)
3. Bertsimas, D., Tsitsiklis, J.N.: Introduction to Linear Optimization. Nashua, NH, (1997)
4. Blum, A., Dunagan, J.: Smoothed analysis of the perceptron algorithm for linear programming. In: Proceedings of the 13th Annual ACM-SIAM Symposium on Discrete Algorithms, publisher: SIAM, Philadalphia, PA, pp. 905–914 (2002)
5. Blum, A., Frieze, A., Kannan, R., Vempala, S.: A polynomial-time algorithm for learning noisy linear threshold functions. Algorithmica **22**(1), (1998)
6. Cheung, D., Cucker, F.: A new condition number for linear programming. Math. Program. **91**, 163–174 (2001)
7. Dantzig, G.B.: Converting a converging algorithm into a polynomially bounded algorithm. Technical Report SOL 91-5, Stanford University, (1991)
8. Dantzig, G.B.: Bracketing to speed convergence illustrated on the von Neumann algorithm for finding a feasible solution to a linear program with a convexity constraint. Technical Report SOL 92-6, Stanford University, (1992)
9. Dantzig, G.B.: An $\epsilon$-precise feasible solution to a linear program with a convexity constraint in $1/\epsilon^2$ iterations independent of problem size. Technical Report SOL 92-5, Stanford University, (1992)
10. Dunagan, J., Vempala, S.: A simple polynomial-time rescaling algorithm for solving linear programs. In: Proceedings of STOC'04, pp. 315–320. ACM Press, New York, NY, (2004)
11. Epelman, M.A., Freund, R.M.: Condition number complexity of an elementary algorithm for resolving a conic linear system. Operations Research Center, Massachusetts Institute of Technology. Working paper, 319–97 (1997)
12. Epelman, M.A., Freund, R.M.: Condition number complexity of an elementary algorithm for computing a reliable solution of a conic linear system. Math. Program. **88**, 451–485 (2000)
13. Frank, M., Wolfe, P.: An algorithm for quadratic programming. Naval Res. Logist. Q. **3**, 95–110 (1956)
14. Freund, R.M.: Dual gauge programs, with applications to quadratic programming and the minimum-norm problem. Math. Program. **38**, 47–67 (1987)
15. Freund, R.M., Vera, J.R.: Some characterizations and properties of the "distance to ill-posedness" and the condition measure of a conic linear system. Math. Program. **86**(2), 225–260 (1999)

16. Freund, R.M., Vera, J.R.: Equivalence of convex problem geometry and computational complexity in the separation oracle model. Math. Oper. Res. **34**, 869–879 (2009)
17. Gonçalves, J.P.M.: A family of linear programming algorithms based on the von Neumann algorithm. PhD Thesis, Department of Industrial and Systems Engineering, Lehigh University, Bethlehem (2004)
18. Khachian, L.G.: A polynomial algorithm for linear programming. Sov. Math. Dokl. **20**, 191–194 (1979)
19. Klafszky, E., Terlaky, T.: On the ellipsoid method. Rad. Math. **8**, 269–280 (1992)
20. Li, D.: On rescaling algorithms for linear optimization, PhD Proposal, Revised. Department of Industrial and Systems Engineering, Lehigh University, Bethlehem, (2011)
21. Minsky, M., Papert, S.A.: Perceptrons: An Introduction To Computational Geometry. MIT Press, Cambridge, MA, (1969)
22. Renegar, J.: Linear programming, complexity theory and elementary functional analysis. Technical Report 1090, School of Operations Research and Industrial Engineering College of Engineering, Cornell University, Ithaca, NY, USA, (1994)
23. Renegar, J.: Some perturbation theory for linear programming. Math. Program. **65**, 73–91 (1994)
24. Roos, C., Terlaky, T., Vial, J.-P.: Interior Point Methods for Linear Optimization. Springer, New York, NY, (2006)
25. Rosenblatt, F.: The perceptron – a perceiving and recognizing automaton. Technical Report 85-460-1, Cornell Aeronautical Laboratory, Ithaca, NY, USA, (1957)
26. Schrijver, A.: Theory of Linear and Integer Programming. Wiley, Hoboken, NJ, (1998)
27. Shawe-Taylor, J., Cristianini, N.: Support Vector Machines and Other Kernel-Based Learning Methods. Cambridge University Press Cambridge, England, (2000)
28. Soheili, N., Peña, J.: A primal-dual smooth perceptron-von Neumann algorithm. In: Discrete Geometry and Optimization Fields Institute Communications **69**, 303–320 (2013)
29. Soheili, N., Peña, J.: A smooth perceptron algorithm. SIAM J. Optim. **22**(2), 728–737 (2012)
30. Todd, M.J., Ye, Y.: Approximate Farkas Lemmas and stopping rules for iterative infeasible-point algorithms for linear programming. Math. Program. **81**, 1–21 (1998)

Printed in the United States
By Bookmasters